THE
CITIZEN-
POWERED
ENERGY
HANDBOOK

THE
CITIZEN-
POWERED
ENERGY
HANDBOOK
COMMUNITY
SOLUTIONS TO A
GLOBAL CRISIS
GREG PAHL

Foreword by RICHARD HEINBERG

CHELSEA GREEN PUBLISHING COMPANY
WHITE RIVER JUNCTION, VERMONT

Editor: Ben Watson
Copy Editor: Cannon Labrie
Proofreader: Eric Raetz
Indexer: Marc Schaefer
Designer: Peter Holm, Sterling Hill Productions
Design Assistant: Daria Hoak, Sterling Hill Productions

Printed in the United States of America
First printing, January 2007
10 9 8 7 6 5 4 3 2 1

Our Commitment to Green Publishing

Chelsea Green sees publishing as a tool for cultural change and ecological stewardship. We strive to align our book manufacturing practices with our editorial mission and to reduce the impact of our business enterprise on the environment. We print our books and catalogs on chlorine-free recycled paper, using soy-based inks whenever possible. This book may cost slightly more because we use recycled paper, and we hope you'll agree that it's worth it. Chelsea Green is a member of the Green Press Initiative (www.greenpressinitiative.org), a nonprofit coalition of publishers, manufacturers, and authors working to protect the world's endangered forests and conserve natural resources.

The Citizen-Powered Energy Handbook was printed on Enviro100, a 100 percent post-consumer-waste recycled, old-growth-forest-free paper supplied by Malloy, Inc.

Library of Congress Cataloging-in-Publication Data

Pahl, Greg.
    The Citizen-powered energy handbook : community solutions to a global crisis / Greg Pahl ; foreword by Richard Heinberg.
        p. cm.
    Includes bibliographical references and index.
    ISBN-13: 978-1-933392-12-7
    ISBN-10: 1-933392-12-6
 1. Power resources. 2. Renewable energy sources. I. Title.

    TJ163.2.P33 2007
    333.79—dc22

                                            2006032425

Chelsea Green Publishing Company
P.O. Box 428
White River Junction, VT 05001
(800) 639-4099
www.chelseagreen.com

This book is dedicated to Dr. Marion King Hubbert, who knew what he was talking about—and didn't give up—even when nobody believed him.

"America is addicted to oil."
—George W. Bush

CONTENTS

7. GEOTHERMAL   225

High-Temperature Geothermal, 225 • Low-Temperature Geothermal, 242
Final Thoughts, 258

8. THE COMMUNITY SOLUTION   259

The Big Picture, 262 • Community-Supported Energy, 267 • Groups
That Get It, 269 • Barriers to Progress, 287 • Changing the Model,
Changing the Debate, 291 • The Wrong Solutions, 292 • Security and
Opportunity, 294

ACKNOWLEDGMENTS

While working on this project, I met and spoke with many dedicated and enthusiastic people who are committed to helping the global community free itself from its dependency on fossil fuels, and who wish to build a better, more sustainable society at the community level. I would like to thank the following, without whom this book would not have been possible: Alan Kurotori, Municipal Solar Utility, City of Santa Clara, California; Nicolas Ponzio, Vermont Solar Engineering, Burlington, Vermont; Elena Kann and Bill Fleming, codevelopers, Westwood, Asheville, North Carolina; Dr. Jason Bradford, organizer, Willits Economic Localization (WELL) project, and Brian Corzilius, Willits Energy Committee, Willits, California; Dan Pellegrini, president, Cooperative Community Energy, San Rafael, California; Steve Lyons, Sine Electric, Santa Rosa, California; Johnny Weiss, executive director, Solar Energy International, Carbondale, Colorado; Soozie Lindbloom, coordinator, Solar in the Schools program, Solar Energy International, Carbondale, Colorado; Dave Borton, Brunswick, New York; Dr. Thomas Reed, Golden, Colorado; Dan Juhl, DanMar & Associates, Woodstock, Minnesota; Jennifer Grove, Northwest Sustainable Energy for Economic Development, Seattle, Washington; Stewart Russell, WindShare; and Doug Fyfe, general manager, Countryside Energy Co-operative, Milverton, Ontario, Canada.

I also want to thank John Warshow, Winooski One Partnership, Montpelier, Vermont; Daniel New, president, Canyon Hydro, Deming, Washington; Fred Ayer, executive director, Low Impact Hydropower Institute, Portland, Maine; Carol Ellinghouse, water resources coordinator, Boulder, Colorado; Avram Patt, general manager, Washington Electric Cooperative, East Montpelier, Vermont; Paul Cunningham, principal, Energy Systems & Design, Sussex, New Brunswick, Canada; Shawn Swartz, Earthaven Ecovillage, Black Mountain, North Carolina; Dan White, editor and publisher, *Ocean News & Technology* magazine, Palm City, Florida; Clare Brodie, Ocean Power Delivery Ltd., Edinburgh, UK; Margaret Murphy, Nova Scotia Power, Halifax, Nova Scotia, Canada; David Brynn, director, Green Forestry Education Initiative, University of Vermont, Burlington, Vermont; Robert Rizzo, director of facilities administration, Mount Wachusett Community College, Gardner, Massachusetts; John Irving, plant manager, McNeil Generating Station, Burlington Electric Department, Burlington, Vermont; Robert Foster, Weybridge, Vermont; Mike Burns, vice president of operations, Market Street Energy, St. Paul, Minnesota; and Carl Lilliehöök and Peter Undén, Svensk Biogas, Linköping, Sweden.

In addition, thanks to Bill Lee, general manager, Chippewa Valley Ethanol Company, LLC, Benson, Minnesota; George Douglas, media relations manager, National Bioenergy Center, National Renewable Energy Laboratory (NREL), Golden, Colorado; Maria "Mark" Alovert, San Rafael, California; Lyle Estill, "cofounder and vice president of stuff," Piedmont Biofuels Cooperative, Pittsboro, North Carolina; Ralph and Lisa Turner, co-owners, Laughing Stock Farm, Freeport, Maine; Jeff Ball, city manager, Klamath Falls, Oregon; Tonya "Toni" Boyd, Geo-Heat Center, Klamath Falls, Oregon; Gerald Nix, technology manager, geothermal and industrial technologies programs, NREL, Golden, Colorado; Bernie Karl (owner) and Gwen Holdmann, Chena Hot Springs, Alaska; Harold Rist II, owner, Smart-Energy, Queensbury, New York; George Hagerty, Queensbury, New York; Jerry Johnson, director of public information, Luther College, Decorah, Iowa; Todd Chambers, utilities director, Pierre, South Dakota; Phil Nichols, Phil Nichols Associates, Rapid City, South Dakota; Angela Sacco, Enwave Energy Corporation, Toronto, Ontario, Canada; Megan Quinn, outreach director, Community Service Inc., Yellow Springs, Ohio; Celine Rich and Julian Darley, cofounders, Post Carbon Institute, Vancouver, British Columbia, Canada; Annie Dunn Watson, cofounder, Vermont Peak Oil Network, Essex, Vermont; and Netaka White, cofounder, Addison County Relocalization Network, Salisbury, Vermont.

Special thanks to Paul Gipe, Tehachapi, California, for his comments and suggestions as well as permission to use excerpts from his excellent book, *Wind Power: Renewable Energy for Home, Farm, and Business*, which were adapted for inclusion in chapter 3, "Wind Power."

I would also like to offer my heartfelt thanks to all the folks at Chelsea Green Publishing. In particular I want to thank John Barstow, Ben Watson, and Marcy Brant, my editors, who assisted me along the way. And I especially want to thank Margo Baldwin for her strong support for this project.

I also want to thank anyone else I may have forgotten to mention here. All of your contributions, both large and small, are greatly appreciated.

Finally, I want to thank my wife, Joy, for her help in proofreading and generally putting up with me while I was trying to bring this project to completion.

FOREWORD

Human history has been marked by three decisive energy transitions—the commencement of the use of fire, the invention of agriculture (including the domestication and harnessing of draft animals), and the adoption of fossil fuels. Of these three, the last—a project that began only about two centuries ago—is proving to have had by far the greatest impact on human population and the environment. Indeed, the fossil-fuel revolution has occurred so quickly, and with such overwhelming force, that it may prove to be humankind's undoing.

The coming century will see the fourth great energy transition—one way or another. This is a certainty; all that remains to be revealed is whether this next transition is undertaken with planning and cooperative effort, or whether it is put off as long as possible. In the latter case, our prospects are not good.

We know that the fossil-fueled energy regime is coming to an end, and that something else will follow, for two reasons.

First, the burning of coal, oil, and natural gas produces carbon dioxide in sufficient amounts to destabilize the global climate. If we do not stop burning these biochemical accumulations of ancient sunlight, we are likely to set off a series of chain reactions (including the melting of polar ice and the release of methane from arctic permafrost) that could turn a mere warming of the globe by a couple of degrees into a full-blown global inferno of climatic chaos, complete with widespread desertification and sea-level increases of over thirty feet or more over the course of the next few decades. In short, a failure to foreswear fossil fuels is likely to be suicidal to our species.

Second, we face the specter of oil and gas depletion. Fossil fuels are non-renewable and therefore finite in quantity, and there is abundant, persuasive evidence that the rate of global oil extraction will peak by the end of the decade and begin its inevitable, inexorable decline. The global natural gas peak will not lag far behind. This might be seen as a good thing, given the climatic consequences of continuing to burn these fuels. However, modern societies have become overwhelmingly dependent on oil for transportation and agriculture, on gas for home heating, power generation, and fertilizer production, and on both as feedstocks for chemicals and plastics. The contemporary urban environment is unimaginable without these services and materials. And so oil and gas depletion threaten both our economy and our way of life. It is not mere scare mongering to draw parallels between the vulnerabilities of modern oil-dependent societies to petroleum depletion, and the circumstances that led to the collapse of ancient civilizations.

In light of these realities, it is clear that the work of engaging proactively, purposefully, and intelligently in the energy transition ahead (that is, the transition *away from* fossil fuels and *toward* renewable alternative forms of energy) represents the most important work of the new century.

There are endless human concerns, many of which are more or less perennial—the quests for justice, equality, knowledge, artistic excellence, personal success, and so on. All of these will and should continue to have importance in people's lives. However unless our species successfully manages the next energy transition, efforts in these directions will have only transitory meaning; in the end, they will be overwhelmed by the disintegration of the social and ecological systems that make human existence possible. On the other hand, if the energy transition is handled well, many of our current social, economic, and environmental quandaries will be substantially ameliorated. Our descendants could live in smaller communities, enjoying a more stable environment as well as more intergenerational solidarity and a deeper sense of connection with place and vocation.

The renewable-energy transition will not happen automatically. If we simply sit back and wait for market forces to propel the shift, the results will be horrific—energy wars, global economic collapse, climatic catastrophe, and widespread, persistent famine.

The situation would look much brighter if there were signs that government and industry leaders already understand the situation and are preparing enormous investments in new energy sources, more efficient transportation networks, and more resilient, organic, locally-based agricultural systems. There are indeed such signs on a very small scale and in the case of a very few municipalities and two or three nations. However, in Washington, Detroit, and Beijing, the predominant discourse is merely about maintaining current trajectories of energy production and consumption.

Therefore much hinges on large numbers of citizens taking matters into their own hands. I am not proposing vigilante action to sabotage oil fields or install guerilla wind turbines. But I am suggesting that widespread, voluntary, proactive efforts by citizens and small communities to ditch fossil fuels and develop alternatives could play an important role in helping society as a whole begin moving in a direction essential to its own survival.

Such efforts must begin with energy literacy. In past decades, it seemed that only techno-nerds were interested in measuring energy efficiency, in understanding how a photovoltaic cell works, or in keeping up with the literature on energy profit ratios. Now it is clear that civilization will persist or perish by our attention to such information. We can no longer afford to leave these essential matters to the energy geeks (bless their hearts). We may not all have the capacity to become experts, but we must all make the effort to understand the basics. Our society must soon and quickly make fateful choices regarding a confusing array of alternative energy sources and conservation strategies, and, because of the late hour and limited budgets, we cannot afford to make many costly, time-consuming errors. The only way to avoid such errors is to invest the effort to understand as much as we can about energy itself, the various available energy sources, the necessary ways of assessing them, and the uses to which we put those resources.

Greg Pahl has done us an enormous favor by assembling a great deal of such information in this highly readable and well-organized book. This citizen handbook provides a valuable overview not only of the various energy alternatives, but also of what is currently being done with them in several nations, and in towns and cities across the U.S. There is useful data here for experts, but it is presented in a format that will primarily benefit the layperson who knows relatively little about energy and wants to learn.

The reader will come away not only better informed, but also better prepared to take action—which is really the point of the exercise. Reducing fossil fuel consumption takes knowledge, planning, investment, and effort. And only those who have successfully reduced their consumption will be in position to show others the way.

To his credit, Pahl maintains an upbeat, encouraging tone throughout this narrative. There are some who look at the implications of our current societal addiction to fossil fuels and conclude that there is no hope. But in fact, no matter how great the challenge, we will all be much better off if we do what we can rather than simply sinking into cynicism and despair. We need to understand the enormity of the task, but we also need courage and good cheer as we apply ourselves to it. And there are plenty of encouraging community energy efforts—in places like Willits, California, Toronto, Ontario, and Burlington, Vermont—to savor and learn from.

This book is for pioneers in the great project of the new century. It will remain a touchstone informational resource for many years to come. May you benefit from it by applying its insights and suggestions in your life and in efforts you undertake with others in your community. May we all engage deliberately, proactively, and enthusiastically in the energy transition, and learn to enjoy life without fossil fuels.

RICHARD HEINBERG
Santa Rosa, CA
October 2006

INTRODUCTION

It's dark and cold outside. Up on the bridge, the captain has been warned about the presence of icebergs, yet the ship is steaming full speed ahead through the North Atlantic. But it's okay. The *Titanic* is unsinkable. And the band plays on.

Nearly one hundred years after the *Titanic*'s ill-fated voyage, we are also embarked on a collision course with disaster, perhaps not as dramatic, but far more catastrophic and far-reaching in its implications. The global economy in general—and the United States in particular—is about to run into an iceberg, despite repeated warnings. The captain is arrogant and inattentive. And, like the passengers on the *Titanic*, the vast majority of the general public is not prepared for the extreme danger that lies dead ahead. When this disaster strikes and everybody rushes up on deck, they are going to find that—even after all these years—there still aren't enough lifeboats.

THE DANGER OF PEAK OIL

The disaster I'm referring to is the onset of the global oil production peak—often referred to as "peak oil" or "Hubbert's Peak" after the Shell geologist Dr. Marion King Hubbert. In 1956, Hubbert accurately predicted that U.S. domestic oil production would reach its peak in 1970. He also predicted global production would peak sometime between 1995 and 2000, which it would have if the politically motivated oil shocks of the 1970s had not delayed the peak for about ten to fifteen years. Generally ignored until just a few years ago, *peak oil* is now viewed by a growing number of observers as a greater (or at least more immediate) danger to human society than global warming. That doesn't mean that global warming isn't a serious problem. It is, especially in the long term. But peak oil is such a threat because the modern global economy is now almost totally dependent on enormous quantities of relatively cheap petroleum products. Anything that seriously disrupts the supply or price of oil means big trouble, and the current volatility of oil prices is already causing problems that are beginning to ripple through the global economy.

So what does peak oil really mean? When we arrive at peak oil, we will have consumed half of the total global reserves of oil. That might not seem like such a big deal, since half the total reserves are still available. But what many people don't realize is that the first half was the easy part to find and exploit, and also represented the highest quality. While the remaining half is still in the ground, it's generally much lower in quality and located in much smaller fields in inconvenient places like the Arctic, or under deep water. Consequently, what remains is going to be much harder and more expensive to produce. In 2003, for example, the global oil industry invested $8 billion in exploration, but only found $4 billion worth of oil. Since then, things have only gotten worse. Poor results like this do not provide much incentive to increase investments in exploration. At some point, it may become uneconomic to extract the remaining oil, and much of it will probably never be recovered. This eliminates an awful lot of the other half of the reserves that people have been counting on to carry us well into the twenty-first century.

Worse yet, serious doubts have recently arisen concerning the size of reserves claimed by many oil companies and oil-producing nations. The massive 2004 accounting scandal involving oil giant Royal Dutch/Shell and the subsequent 22 percent (4.35 *billion* barrels) cut in the company's petroleum reserve estimates is viewed by some industry experts as just the tip of the iceberg of overinflated reserve figures for the industry. Matthew Simmons raises similar concerns about Saudi Arabian reserves in his 2005 book, *Twilight in the Desert: The Coming Saudi Oil Shock and the World Economy*. Simmons maintains (over the loud protestations of the Saudi government) that the commonly held assumption that Saudi Arabia will be able to continue to produce oil at current levels for the next fifty years is unrealistic. He thinks Saudi production has *already* peaked or is about to peak. Simmons, the highly respected founder and chairman of the world's largest energy investment banking company, makes a detailed and convincing case for his assessment. Not everyone agrees with Simmons, but in

March 2005 the Algerian minister for energy and mines admitted that OPEC has essentially reached its oil production limits.[1] And in April 2006 United Arab Emirates' oil minister Mohammed bin Dhaen al-Hamli confirmed this when he said "fundamentally, there is nothing we can do" about current high oil prices.[2]

Adding to these concerns, in January 2006, the Kuwait Oil Company admitted that its supergiant Burgan oil field has peaked. And in February, a leaked internal memo from Mexico's state-owned oil company, Petroleos Mexicanos (Pemex), disclosed that the world's second-largest oil field, Cantarell, has also peaked. "It's a supergiant field, so when you have a supergiant field declining, it's very difficult to compensate for that," said Adrian Lajous, a veteran oilman and the director of Pemex from 1995 to 1999. "Cantarell has peaked and has started its decline."[3] Taken together, these reports are not good news.

The main problem is that, after peak production is reached, the global supply of oil will inexorably begin to decline at the same time that demand continues to increase. The global oil market currently consumes 84 million barrels a day. By 2025, that demand is expected to climb to around 121 million barrels a day, according to the U.S. Energy Information Administration. At the same time, production from most of the largest existing oil fields is

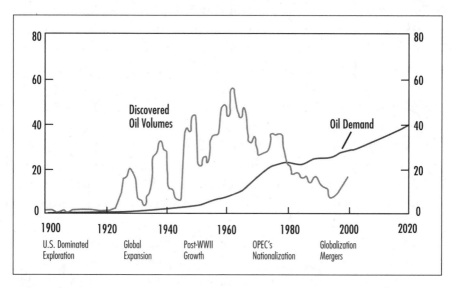

Billions of oil-equivalent barrels. Exxon-Mobil.

declining at a rate of 5 percent or more annually. (Cristophe de Margerie, head of exploration for French oil giant Total says, "Numbers like 120 million barrels per day will never be reached, never.")[4] When the line on the chart for supply that's coming down crosses the line on the chart for demand that's going up, we will have reached the critical tipping point. The huge price increases for oil predicted after we reach the tipping point will unquestionably lead to much higher prices for almost everything—especially food. The remarkable "Green Revolution" that now feeds the world's 6.6 billion inhabitants is based mainly on cheap oil (and natural gas). The end of cheap oil will undoubtedly cause serious disruptions in the production of food, causing food prices to skyrocket. As a result, millions of people, particularly in struggling Third World countries, will almost certainly starve. But hunger may also stalk the streets of cities and suburbs in the developed nations as well. Prices for most other goods will also rise dramatically as we slide further down the backside of Hubbert's Peak and as demand for oil continues to bump up against an irreversible decline in supply.

COMING SOON

Unfortunately, the onset of peak oil may arrive much sooner than most people think. The experts, as always, are divided on when this will take place. One of the most optimistic views, promoted by the U.S. Department of Energy, maintains that oil production won't peak until 2037. Many observers feel this estimate is far too optimistic, especially considering the huge increase in demand from countries like China, which overtook Japan as the world's second-largest oil consumer in 2003. Renowned petroleum geologist Colin Campbell estimates that global extraction of oil will peak sometime before 2010. Kenneth Deffeyes, a geophysicist and author of the 2005 book *Beyond Oil*, says the date for maximum production was December 2005. "My career as a prophet is over, I'm now a historian," he said.

But the exact date may be somewhat academic. This is because, if the rollover point at the top of the peak does occur within the next few years, there simply isn't enough time left to make the massive shift to the renewable energy strategies that would allow for a smooth transition from our present oil-based economy. That smooth transition would have been possible if we had started the process twenty or thirty years ago (as many environmentalists and scientists urged at that time). But it didn't happen. Now, we're simply not prepared, and the consequences will almost certainly be catastrophic. A February 8, 2005, report prepared for the U.S. Department of Energy (commonly referred to as the "Hirsch Report") entitled *The Mitigation of the Peaking of World Oil Production* puts it this way: "The world has never faced a problem like this. Without massive mitigation more than a decade before the fact, the problem will be pervasive and will not be temporary. Previous energy transitions were gradual and evolutionary. Oil peaking will be abrupt and revolutionary."[5]

Unfortunately, we are also facing a similar dilemma with natural gas, which was supposed to be a cleaner and abundant substitute for oil and coal. It turns out that domestic supplies of natural gas are declining almost as rapidly as supplies of oil. And filling the gap with foreign imports of liquefied natural gas (LNG) is, at best, a problematic—and only temporary—strategy, since natural gas (another fossil fuel) is part of the problem rather than a long-term solution, according to Julian Darley in his 2004 book, *High Noon for Natural Gas: The New Energy Crisis*. "The speed and suddenness with which the natural gas crisis appears to have arrived in North America is due in large part to the silence of the extraction industry, but it is also abetted by the myth of inexhaustibility and substitutability of resources," he says in the book.[6]

It's also likely that this crisis will be arriving sooner than expected in Europe as well because, on April 20, 2006, Gazprom, the Russian energy giant, admitted that it had significantly less natural gas in its reserves than previously assumed, and that in a few years' time Russia would not be able to meet the demands of all its consumer countries. This has enormous implications for the European as well as Asian nations that have been counting on Russian natural gas supplies.[7] At the very least, considering the impending reductions in the supply of oil and natural gas, the global economy is about to come in for a very hard landing. A series of global economic recessions or depressions, massive unemployment, political instability, and more international conflict are almost certain to follow.

In addition to these oil and gas woes, we are simultaneously facing many other challenges. Global warming, or as environmentalist David Orr aptly describes it, "global destabilization," is right at the top of the list, and its effects are becoming harder and harder to ignore. This unfortunate side effect of our profligate burning of fossil fuels is already beginning to wreak havoc around the world. The recent record floods and disastrous wildfires in nations throughout the globe and the massive devastation caused by four major hurricanes in Florida in 2004—and especially hurricanes Katrina, Rita, and Wilma in 2005—are just a preview of what's to come. I'm sorry to say that the city of New Orleans was the first major casualty of global warming in the United States. The second major casualty was the U.S. oil and gas industry in the Gulf region, which may never recover fully. These events, more than anything, have provided a loud wake-up call, but many Americans still don't really get it. There will be similar events in the very near future, however, that will be impossible to ignore. What's more, the string of recent record-high temperatures—2005 was the warmest year since records were kept, according to NASA—does not bode well for the future either. A June 2006 National Academy of Sciences report to Congress says that "recent warmth is unprecedented for at least the last 400 years and potentially the last several millennia . . . and that human activities are responsible for much of the recent activity."

A large body of published scientific studies in 2005 simply confirmed what most responsible scientists have been saying for years, that the physical consequences of climate change are no longer theoretical, they are real, they have arrived, and they can be quantified. Taken collectively, these studies suggest that the world may well have moved past a crucial tipping point. What's more, the science behind these studies also makes it clear that additional climate effects will result "even if emissions of greenhouse gases are halted immediately."[8] Halted immediately? About the only thing we can be certain about is that greenhouse gas emissions are *not* going to be halted any time soon, especially considering the intransigent position taken by the Bush administration on the whole subject of global warming, and its plans to rely even more heavily on coal in the future.

In the first months of 2006, as growing evidence of a dramatic acceleration of global warming proliferated around the world (but especially at the poles), the tone of the urgent warnings from climate scientists became even more desperate. But it's no longer necessary to listen to scientists to get the message. Just look out the window—but beware of what you might see. The more than 100 killer tornadoes that struck five Midwestern states in a two-day period in-mid March 2006—storms that killed at least ten people and caused as much damage in forty-eight hours as might be expected in an entire year in some locations—added even more evidence that something is terribly wrong with global weather patterns. That was followed by 63 devastating tornadoes—some with winds over 200 miles per hour—in seven states on April 3 that killed at least twenty-seven people and totally obliterated hundreds of homes and businesses throughout the region. The very next week, twelve more people died as numerous twisters struck Tennessee. And what were described as "300-year floods" on the East Coast in late June 2006 took at least twenty lives in Pennsylvania, Maryland, New York, and Virginia. The "killer heat wave" of July that gripped virtually the entire nation added an exclamation point to this ominous trend. This is a clear example of why global warming is so important. We're not just talking about a gradual adaptation to warmer temperatures, or about being able to grow bananas in Vermont someday.

The combined impact of global warming, peak oil, and overpopulation, along with the alarming depletion of freshwater resources and habitat destruction around the world, is setting the stage for a global catastrophe. We are facing challenges of biblical proportions. It's increasingly possible that industrialized society as we know it will collapse under the combined, simultaneous pressures of these many challenges, and a growing number of observers are warning of this potential.

PROBLEM? WHAT PROBLEM?

Unfortunately, our national political leaders, especially in Washington, D.C., are in denial. The passage of the controversial Energy Policy Act by Congress in July 2005 is a perfect example. The energy bill was a golden opportunity to substantially change the direction of the nation's energy policies for the next decade or so. Yet, for the most part, it was simply business as usual. While about 25 percent of the total funding represented in the bill was allocated to a variety of token renewable energy and conservation initiatives, the lion's share—65 percent—went to subsidize the oil, natural gas, coal, and nuclear industries. Giving subsidies for oil and natural gas companies at the same time they were raking in billions of dollars in record-high profits was simply obscene, to say nothing of a waste of taxpayer money that should have been invested in renewables. What's more, a provision for higher vehicle fuel-efficiency standards was stripped from the bill, eliminating a key tool for reducing oil consumption nationwide. (Higher fuel-efficiency standards for some vehicles proposed later by the Bush administration didn't even begin to address this problem and were mostly just window dressing.) A target of producing 10 percent of the nation's electricity from renewable sources by 2010 was also removed from the bill at the last minute. The biggest outrage of all is that the members of Congress who passed this do-nothing bill actually get paid for this sort of head-in-the-sand behavior.

One noteworthy exception to this sorry state of affairs inside the Beltway is Republican Congressman Roscoe Bartlett from Maryland. In early 2005 Bartlett got the peak oil message, and he spent the next year trying to get the attention of his congressional colleagues—and the rest of the nation. At first, his colleagues could barely suppress their amusement, thinking that Bartlett was committing political suicide. But due at least in part to his tireless efforts (and the efforts of members of the grassroots peak oil community), by early 2006, peak oil was finally the subject of at least some open discussion in the nation's capital. Congressman Bartlett deserves credit for political courage in a place conspicuously lacking in same.

It is increasingly clear from polls conducted in March of 2006 that the national mood about energy matters at the grassroots level has shifted dramatically, and that the politicians inside the Beltway, but especially President Bush and his administration, are increasingly out of step with the rest of the nation. More than four out of five Americans (83 percent) support "more leadership from the federal government to reduce the pollution linked to global warming, encourage new approaches to promoting conservation, and spark the development of renewable or alternative energy sources." The level of support was relatively uniform across political lines, including 81 percent of conservatives, 83 percent of independents, and 88 percent of liberals.[9]

Many cities and states across the nation are also fed up with the lack of action inside the Beltway on global warming, and have essentially bypassed both Bush and Congress to tackle the problem themselves. In May 2005, the mayors of more than 130 cities, including New York and Los Angeles, agreed to meet the emissions reductions envisaged in the Kyoto Accord, independent of federal policy decided in Washington. The number of cities has now climbed to 238, and the mayors, representing over 44 million people, have signed an agreement to meet the goals spelled out in the 1997 international treaty on climate change and urged the federal government to do the same.

Unfortunately, the mainstream national "news" media in the United States—especially commercial TV news—has all but abdicated its responsibility to inform the electorate about the important issues of the day. As a result, until very recently the vast majority of the general public in the United States has been almost totally clueless about the disaster that is about to overtake them. This disaster, appropriately named "the Long Emergency" by James Howard Kunstler in his 2005 book of the same title, is going to change everything. At best, we will have to rethink many of our basic assumptions about our economy and society, and be forced to do things differently. At worst, our modern technological civilization may come to a terrible and tragic end, according to Kunstler.

But how is it that what is arguably the most serious challenge to our very survival has been almost totally ignored (or deliberately suppressed) by the people who are supposed to be looking out for our interests? The main problem is that they are too busy looking out for their own interests or the interests of the powerful corporations they represent. Bad news doesn't inspire people to go shopping at Wal-Mart, or to invest on Wall Street, which largely explains the longtime blackout of coverage in the mainstream broadcast media. And what politician do you know who has the courage to stand up in front of national TV cameras and tell the electorate that the American Dream is about to come to an end?

The Great Wall of Silence finally began to crumble in late 2005, as articles about peak oil began to appear in a few mainstream print publications such as *Time* magazine and *USA Today*. Then, on March 1, 2006, *The New York Times* finally said in an editorial, "The concept of Peak Oil has not been widely written about. But people are talking about it now. It deserves a careful look—largely because it is almost certainly correct." Better late than never (Al Gore's powerful 2006 documentary, *An Inconvenient Truth*, has finally helped to move the global warming discussion into the mainstream as well).

THE RESPONSES

There are basically four main responses to this impending emergency. Richard Heinberg describes them in his 2004 book, *Power Down: Options and Actions for a Post-Carbon World*. The first response is based on fear, division, competition, and conflict. Heinberg refers to this strategy as the "Last One Standing." Human history simply overflows with examples of wars fought over critical resources, especially dwindling resources. Many observers point to the Bush administration's foreign policies since the 2000 presidential election as a classic example. They describe the unfortunate and disastrous 2003 U.S. invasion of Iraq as the opening salvo of the coming global oil wars. And many of those same observers say that China will be a key competitor in that struggle in the years ahead. But if the scramble to control the remaining global oil and gas reserves descends into international conflict, there will be many contestants vying to be the last one standing (India's oil consumption is predicted to increase nearly 30 percent in the next five years). But by then it won't matter much, since humanity will have squandered its precious dwindling resources on warfare and destruction that could have been used for more productive purposes.

Although the long-term consequences of this strategy are bleak, many national leaders already seem committed to following this ultimately self-destructive path. It's more politically expedient to find scapegoats and demonize others than it is to face up to the problems at home and suggest conservation strategies and changes in the lifestyles to which we have become accustomed. Vice President Dick Cheney's statement that "the American way of life is not negotiable" is a classic example of this intransigent attitude. We especially need to be wary of politicians who will try to divide us and use fear to divert public attention away from their own failures to adequately prepare the nation for this disaster. The opportunities for political mischief are enormous.

The second response to the Long Emergency is denial and wishful thinking. Heinberg describes this as "Waiting for the Magic Elixir." There is a good deal of both denial *and* wishful thinking on the part of many government leaders, economists, and market analysts (as well as the general population) when they are confronted with our many upcoming energy challenges. "Don't worry, our scientists will think of something," is a frequent response. "All we have to do is just . . . ," is another (where you simply fill in the blank with your favorite silver-bullet energy strategy). "The free market will solve everything," is yet another. Right.

Many folks just can't seem to cope with the idea that our sophisticated, technological, modern way of life could possibly be imperiled, and dismiss the threat out of hand, or just hope the problem will go away. Some don't want to listen. Some are delusional. At noon, on April 27, 2006, in response to high gasoline prices, various Christian clergy from around the nation converged on a gas station in Washington, D.C., to pray for lower prices. In a press release, the Pray Live group said many people are "overlooking the power of prayer when it comes to resolving this energy crisis."[10] But all the wishful thinking in the world isn't going to solve this problem. At best, it offers false hope. At worst, it insures disaster. Meanwhile, the Long Emergency looms ever closer.

The third response to the Long Emergency involves facing the problem squarely, overcoming fear, building community, and following "The Path of Self-Limitation, Cooperation, and Sharing," as Heinberg describes it. Ideally this should be the strategy of choice. But from a political standpoint, expecting national leaders to suggest anything even faintly resembling this strategy is probably unrealistic. One can always hope, but in all likelihood, this isn't going to happen. So, in the absence of leadership at the international or national level, what can we do?

This brings us to the fourth response to the Long Emergency, "Building Lifeboats: The Path of Community Solidarity and Preservation." It's increasingly clear that we cannot wait any longer for our government to solve these problems or wait for some "miracle" technological fix. It's time to get active at the local level. We need to build lifeboats—and get ready to use them. Thanks to the criminal negligence of our leaders, it is almost certainly going to devolve on individuals and communities to fend for themselves, especially if the interconnected global economy—which is hopelessly dependent on cheap oil—collapses.

The intelligent response is to begin the process of localizing and decentralizing our economy now, while we still have the time and resources to do it. And we need to accomplish this local lifeboat-building process as soon as possible. Irish economist and author Richard Douthwaite focuses on this local approach in his excellent 1996 book, *Short Circuit: Strengthening Local*

Economies for Security in an Unstable World (now, unfortunately, out of print and very hard to find). "A community wishing to minimize the hardships it would suffer if the world financial system collapsed should obviously make monetary independence its first priority. A currency and banking system that can continue to serve a particular area regardless of whatever financial convulsions take place outside that area is fundamental to the construction of a self-reliant local economy," he says. "Once a local financial system is in place, the community should turn its attention to meeting its irreducible energy, food and clothing needs from its own area. In fact, I rate community energy independence second only in importance to monetary independence because food production and many other activities depend on energy use."[11] We will be focusing on local energy independence in this book.

THE ENERGY DILEMMA

But you don't have to sit in the dark and cold while trying to deal with all of this by yourself. That's where a copy of *Citizen-Powered Energy Handbook: Community Solutions to a Global Crisis* is going to come in handy. Although there are many problems that we will need to face in the difficult times ahead, one of the most immediate is our current energy dilemma, and the need to shift to renewable, non-fossil-fueled sources of clean energy. The time for debate about this necessity is long past. The time for immediate action is now. This book addresses this imperative with a thorough survey of our current energy options. It also offers an upbeat (but not Pollyannaish) practical response to the Long Emergency that begins at the local level. This book should inspire you to take cooperative action now in your own community, to initiate disaster preparedness plans and begin to revamp our local energy infrastructure for greater energy self-sufficiency while we wean ourselves from our addiction to fossil fuels. This will not be easy, but it's still possible.

Having said that, I want to be clear about this—we're no longer talking about business as usual—this is *a real emergency*. The economic chaos and social disruptions caused by the end of cheap oil and other environmental problems are almost certainly going to result in the development of an intensely local focus on our daily lives. Out of necessity, the primary focus is going to be on the basics—food, water, clothing, shelter, security—and the energy to power a much-reduced and decentralized economy. Without energy, however, we face a bleak future indeed. But with an assured supply of locally produced energy, coupled with a major reduction in consumption, we may have the means eventually to refashion our economy and society into a more sustainable model. And the energy sources for this new economy will also have to be sustainable, otherwise we will effectively seal our own fate. Burning more coal is *not* the answer.

The challenges we face are enormous, and consequently require a collaborative response. *The Citizen-Powered Energy Handbook* focuses on what you and your community can do to prepare for the energy crisis that we will be facing in the very near future. The strategies described are based on current, proven technology (rather than theoretical or unproven energy sources) and with a strong emphasis on cooperative community strategies based on local ownership. This local ownership model is increasingly being referred to as community supported energy (CSE), which is similar to community supported agriculture (CSA), except that instead of investing in carrots, tomatoes, or onions, local residents invest in renewable energy projects and a cleaner environment. Local ownership is the key ingredient that transforms what would otherwise be just another corporate energy project into an engine for local economic development and greater energy security. Examples of what people are already doing collaboratively in communities around the world are provided to offer successful models that can be replicated elsewhere. The book is divided into eight chapters:

Chapter 1 provides a detailed look at our current energy choices—fossil fuels, nuclear, and a wide range of renewables coupled with conservation—and lays the groundwork for the more detailed descriptions of the renewable energy strategies contained in the remaining chapters.

Chapter 2 addresses *solar energy*, and describes the many ways it can be used to power and heat the new downsized, localized economy as well as your home.

Chapter 3 focuses on *wind power*, and looks at the many recent developments in wind technology and where it can best be utilized to power your home or your community, with emphasis on community-owned wind projects.

Chapter 4 looks at *water power*, especially small-scale hydropower, and how it may once again play a significant role in some locations to power your home, business, or community. The chapter also describes some recent ocean energy developments that look promising.

Chapter 5 explores our many *biomass* options, and how they will be a significant part of the post-carbon economy to provide heat, electricity, and some transportation.

Chapter 6 focuses on *liquid biofuels* such as biodiesel and bioethanol for heating and transportation, and spells out the considerable potential, and limitations, of this rapidly growing sector.

Chapter 7 covers *geothermal* resources used for electrical generation, heating, and many other purposes on both a small and large scale, and explains why this strategy offers so much additional potential.

Chapter 8 sums up our renewable energy prospects and places them within the context of the growing international peak oil response, known as the "relocalization movement." We'll see how community-supported energy initiatives can be merged seamlessly with this movement to create a powerful new model for local self-reliance and security.

A glossary of terms, a bibliography, and an extensive guide to organizations and online resources rounds out the volume.

Many of the strategies described in this book can be implemented by you in your own home. Some are more appropriate for businesses, neighborhoods, communities, or even cities. All are intended to help free us from our addiction to fossil fuels and provide at least a basic supply of energy for a new, downsized, localized economy. It's important to understand, however, that this new economy is going to involve significant reductions in consumption at all levels and substantial changes to our lifestyles and patterns of living. This will be an extremely painful process for many people. Some view this as a disaster. Others see it as an opportunity to refashion our society into a more sustainable and equitable model based on strong local communities and cooperation. I prefer to see the glass as half full.

THE CHALLENGES AHEAD

I'm not, however, minimizing the challenges ahead, which are enormous. Population—or more to the point, overpopulation—is the largest problem of all, and is driving most of the other difficulties we face. But since it is basically impossible to have a sensible discussion of this taboo subject, even among otherwise intelligent and rational people, it is increasingly unlikely that we will be able to resolve this situation voluntarily. Consequently, Mother Nature is about to resolve it for us involuntarily.

I also freely admit that many of the difficulties we will be confronted with in the energy-constrained decades ahead cannot be solved by installing solar panels, wind turbines, or small-scale hydroelectric systems. The complex challenges of reinventing our agricultural practices, revitalizing the local business sector, revamping the monetary system, reorganizing transport, and transforming our cities, towns—and ourselves—extend well beyond the scope of this book. Nevertheless, these vital tasks will require the same type of courageous, collaborative strategies that I suggest for renewable energy initiatives. But the goal—a transformed, cooperative, and ultimately sustainable society—is worth the effort, and will be a priceless gift to our children and grandchildren.

In April 2006, as I was completing this book, the general upward trend in global oil prices began to accelerate dramatically. On April 17, the price had broken the $70-per-barrel threshold, and five days later the price had surged to a record $75.35 per barrel. The price of gasoline at the pump in the United States climbed rapidly as well, soon topping $3.00 per gallon in many areas. Most mainstream commentators blamed the run-up on concerns about the Bush administration's saber rattling in the ongoing U.S.-Iranian nuclear crisis, as well as worries about the civil unrest that has cut Nigeria's oil production and other anxieties about tight oil supplies and surging demand. All of this was coupled with the onset of the summer driving season and—incongruously—record high oil inventories in the United States. Something strange was happening. Politicians pointed their fingers at the oil industry, and called for more congressional investigations into price gouging, and talk of an excess-profits tax resurfaced. Other observers blamed oil market speculators. Outraged American drivers complained about "ridiculously high" gasoline prices while filling the oversized gas tanks on their SUVs. At the same time, Lee Raymond stepped down from his position as Exxon/Mobil CEO with a retirement package of nearly $400 million, in what Senator Byron Dorgan from North Dakota described as "a shameful display of greed." Most people across the country seemed to agree.

But while all this was going on, at least one Dallas, Texas, petroleum geologist, Jeffrey J. Brown, claimed that the mainstream media and other observers had it wrong. "A careful examination of recent supply data from the U.S. Energy Information Agency (EIA) suggest a different reason—oil importers are bidding against each other for available total petroleum (crude oil plus product) imports," he said. He then went on to detail falling average daily U.S. net petroleum imports compared to the previous year. "This sharp decline in net U.S. petroleum imports corresponded to the beginning of the recent run-up in oil prices," he observed. According to the EIA, December 2005 appears to have been the all-time record high for world crude production. The latest data, for January 2006, show a decline of about 500,000 barrels per day, according to Brown.[12] This may prove to be a temporary anomaly. But maybe not. If not, then Kenneth Deffeyes may have been correct in his prediction that world oil production would peak in December 2005. I hope he's wrong. But if this *is* the case, then we have already left the familiar before peak oil (BPO) world behind, and have entered the new, unfamiliar post peak oil (PPO) world. It may take several years' worth of additional oil production data before we really know for sure. Regardless of whether Deffeyes is right or wrong, the need for immediate action to prepare for (or respond to) peak oil couldn't be more urgent. So much to do. So little time. Let's get to work.

CHAPTER ONE

OUR ENERGY CHOICES

In the summer of 2005, millions of Americans were shocked when they opened their pre-buy letters from their fuel oil suppliers. On average, the price of heating oil for the 2005–06 season jumped a staggering 40 percent over the previous year, from around $1.50 per gallon to about $2.15 per gallon. Many people, especially those on fixed incomes, were left wondering how they were going to be able to afford to heat their homes. Some were faced with the terrible prospect of having to choose between heat and food. And consumers in general had a hard time coming to grips with the idea that fuel oil at more than $2 a gallon in a pre-buy program was somehow a "bargain." (The 2006–07 pre-buy price was between $2.59 and $2.69 per gallon, if you could get it. Many fuel dealers simply did not offer a pre-buy option.)

A short time later, the price for a barrel of oil hit a new record of $62.30, and proceeded to climb to $70.80 in late August 2005 following the damage to U.S. oil production and refining facilities in and around the Gulf of Mexico caused by hurricanes Katrina and Rita. Immediately after the hurricanes, the price of gasoline temporarily hit a record of over $3 per gallon in the United States as strong summer driving demand bumped up against tight supplies. As a result, many American drivers finally began to lose their love affair with SUVs, while U.S. automakers, who have relied on these wasteful cash cows for years, failed once again to fully understand the implications, and continued their suicidal race to insolvency.

And as the price of aviation fuel headed into the stratosphere, airline executives were sounding increasingly desperate. "No airline can make money with $50 crude oil prices; now we are up to $60," Gary Kelly, the CEO of Southwest Airlines said in a June 24, 2005, interview.[1] Since then, the price of oil has broken the $75 per barrel threshold and commercial jet fuel rocketed to $1.81 per gallon in early 2006, a staggering 90 percent increase from 2001. According to the Air Transport Association, it is extremely difficult for airlines to be profitable when the average price for a gallon of jet fuel exceeds the $1.67 threshold.[2] As airline bankruptcies multiply, the so-called legacy commercial air carriers appear to be in a tailspin. And some of the discount airlines may not survive all that much longer either. The high cost of oil is beginning to ripple through other sectors of the economy as well, with air freight, trucking, manufacturing, and big box retailers all beginning to feel the bite.

JUST THE BEGINNING

This upward spiral of oil prices is just the beginning of the peak oil scenario predicted by Dr. Marion King Hubbert back in 1956. It also marks the onset of the early stages of the economic, political, and social distress described by James Howard Kunstler in his 2005 book, *The Long Emergency: Surviving the End of the Oil Age, Climate Change, and Other Converging Catastrophes of the Twenty-first Century*. "Americans are woefully unprepared for the Long Emergency," Kunstler says in his book. "The American way of life—which is now virtually synonymous with suburbia—can run only on reliable supplies of dependably cheap oil and gas. Even mild to moderate deviations in either price or supply will crush our economy and make the logistics of daily life impossible."[3]

The main problem is that the sprawling suburban development patterns in the United States make it virtually impossible to do anything without driving. Even if you just want to get a gallon of milk, you have to drive—often for very long distances. And the support infrastructure for this spread-out, wasteful mess, including roads, electricity, water, sewer, telephone, as well as fire, police, and other emergency services, is very expensive to build and maintain. All of this is grossly energy inefficient. We're in trouble.

But the main problem with peak oil is not that we are about to run out of oil any time soon. We're not. The real problem, as Kunstler and others point out, is that even small reductions of as little as 5 to 10 percent in the supply of oil can cripple an oil-dependent economy (like ours), eventually leading to a total collapse of the system. And anyone who thinks that our sophisticated, technological society is somehow immune from this fate had better think again. There is a long history of prior societies that have collapsed due to environmental and economic crises, especially if they ignored the problem or failed to respond in time. In his 2004 book, *Collapse: How Societies Choose to Fail or Succeed*, Jared Diamond says "even the richest, technologically most advanced societies today face growing environmental and economic problems that should not be underestimated."[4] Unfortunately, at the present time, our society seems to be choosing to fail.

The problem today is that human society is now global, and our world economy is increasingly interdependent—and virtually all of it relies on cheap oil and natural gas. If you remove that basic foundation, the entire house of cards may very well come tumbling down. The more sophisticated and technological the society, the harder the fall. The crash of the U.S. economy could be spectacular—and terrifying. Most Americans simply won't be able to believe that it is happening.

ENERGY SURVIVAL PLAN

The unfortunate folks at the bottom of the economic ladder are going to feel the pain of peak oil first, just the way they bore the brunt of Hurricane Katrina on the Gulf Coast, most especially in New Orleans. But they won't be the only ones; middle-class Americans will be next. The magnitude of the suffering in the United States will depend to a large extent on how doggedly we cling to the lifestyles to which we have become accustomed, and, especially, to the unsustainable suburban model of living that will no longer make any sense. Walking away from all those overpriced McMansions in suburbia is going to be extremely difficult for most people. But staying might be even worse. Suburbia may very well become the slums of tomorrow, according to Kunstler.

Regardless of where you live, however, this suffering may be reduced somewhat if we immediately institute an urgent program of massive energy conservation, coupled with the installation or adoption of every viable form of renewable energy that is currently available at the local level. This probably will not avoid the crash—I think that it's too late for that—but it may soften the blow a bit, and that may make all the difference. But this strategy needs to be coupled with a willingness to be extremely flexible and nimble in our thinking and in our responses to the many problems we will encounter as we inexorably slide down the back side of Hubbert's Peak. We're going to have to reinvent our entire economy at a time when the resources to do so are going to be seriously constrained by the end of cheap oil and the onset of a protracted recession, depression, or—what's more likely—a series of deepening depressions. To meet this challenge, we're going to have to set aside our differences and cooperate with each other as if our lives depended on it, because quite literally they may.

The intelligent response is to begin the process of localizing and decentralizing our economy now while we still have the time and resources to do it. And we need to accomplish this lifeboat-building process as soon as possible. But it's important to understand that the renewable energy initiatives I'm talking about are not going to be able to sustain our present levels of consumption. Forget that. We're talking about emergency energy to help us provide at least basic goods and services for daily survival until we learn to live within the planet's carrying capacity and eventually shift to a more sustainable economy. No more hopping in the car for a Happy Meal. No more jet travel for vacations in exotic foreign lands. No more driving across the country to Disneyland or Las Vegas for a short vacation. This is because the cost of the fuel for these activities will probably be prohibitively expensive and extremely limited in supply. The fuel that *is* available will be conserved for essential services, like food production and space heating. And if the global economy tanks, as seems increasingly likely, most people simply aren't going to have the money to spend on tourism and entertainment anyway. Most of us will probably be trying to figure out where our next meal is going to come from, especially if the U.S. agribusiness sector—which is almost totally dependent on oil and petrochemicals as well as natural-gas-derived nitrogen fertilizers—collapses. Small-scale family farming, especially organic farming, is going to experience spectacular growth in the years ahead. Consider a career change.

OUR ENERGY DILEMMA

In order to implement a viable survival plan, we need to have a better understanding of the dilemma we face. To do that, we need to take a closer look at our present energy situation. The energy sectors we routinely take for granted fall into three main categories: transportation, electrical generation, and space heating (or cooling in hot climates). Let's take a look at each.

Transportation

The sector of our economy that is going to be hit the hardest by peak oil is transportation. This is where we use the largest quantity of oil and where we are the most vulnerable. Transportation accounts for a staggering 95 percent of global oil consumption. The global economy's cars, trucks, buses, airplanes, boats, and some trains run on gasoline, diesel fuel, or jet fuel. Travel by air and car are the two most energy-inefficient methods of transporting people, and (of course) the two preferred strategies in the United States. But boats and trucks play a key (and somewhat less obvious) role in our economy. All those inexpensive manufactured products that come from overseas, especially China, are transported first by cargo-container ship, and then mainly via truck to a big box store near you. How long do you think Wal-Mart will last when transport fuel becomes prohibitively expensive and their oil-fueled, twelve-thousand-mile-long, just-in-time supply chain breaks down? Not long, especially since their customers won't be able to afford to drive to the mall any longer, either. Big box stores are probably doomed, along with the shopping malls that surround them. But that's not all. Virtually every business and most other institutions and organizations depend on cheap oil-based transport as well. How do most workers get to their jobs? How do children and teachers get to school? How does the mail get delivered? How do police and emergency crews provide their services? You get the idea.

Then there's food. The production and transport of the vast majority of food in most countries is also highly dependent on cheap oil. The production of the food in the first place is a huge issue. But once produced, how does the food get from the field to the processor, from the processor to the distributor, and finally from the distributor to the retailer? Mostly by truck. And the supply lines for foreign-produced foods are even longer and more energy consumptive. The fact that the average American meal travels about fifteen hundred miles or more from farm to plate is not at all reassuring. Remember those images of bare food-store shelves in the former

Soviet Union? Be prepared for a similar sight in a supermarket near you, if you can afford to get there. Although the disruption of transport is going to be a global phenomenon, it's going to be most pronounced, and hardest to deal with, in the United States because of our unfortunate patterns of suburban sprawl and the resultant destruction of prime farmland. Virtually every U.S. city is now surrounded by miles and miles of pavement and unproductive suburban tract housing for as far as the eye can see. Where is the food for these cities going to come from without cheap oil transport?

There is, however, one bright spot in the transport sector—railroads. Although most long-haul trains in the United States are pulled by diesel-powered locomotives, they are *far more* efficient on a per-passenger or per-ton basis than trucks and planes. And diesel locomotives could use biodiesel blends as fuel. Better yet, some railroads are still electrified, which means that, on those lines at least, the trains can be hauled by electric locomotives (assuming that the electrical grid remains viable). Electrified, urban, high-performance light-rail transit and streetcars offer a lot of potential as well. What's more, railroad tracks are far less expensive to build and maintain than multilane superhighways. This long-neglected sector of U.S. transport is about to experience a major comeback. Even Amtrak, the long-suffering stepchild of the public-transport system, could become a vital link for millions of people—as long as the Bush administration does not carve it up and destroy it first—something that they have been trying very, very hard to do. "Obviously, what their goal is—and it's been their goal from the beginning—is to liquidate the company," former Amtrak president David Gunn said in an interview after his politically motivated firing in November 2005.[5]

Electrical Generation

If oil makes the wheels of global transport go round, then electricity makes just about everything else in our modern, technological society function, especially in large cities. Without it, virtually everything grinds to a halt. If you've ever been standing in a supermarket checkout line or riding in an elevator during a power outage, you know what I'm talking about. In addition, during a power failure, heating or air-conditioning (depending on the season) shuts down, the contents of freezers start to thaw, traffic lights stop working, subways and electric commuter rail lines stop running, computers fail, equity markets and banking are disrupted, homes and businesses go dark, cell phones stop working, many water systems stop pumping, and sewage treatment plants fail. The list goes on and on. The power blackout that hit North America on August 14, 2003, cost an estimated $6 billion in outage-related financial losses. Consequently, the importance of maintaining the electrical grid is obvious. So just how vulnerable is the electrical grid to peak oil?

In October 2003 the total U.S. net generation of electricity was 304 billion kilowatt-hours (kWh). Of that total, 51 percent was generated by coal, while nuclear power was responsible for 20 percent, natural gas 18 percent, hydroelectric 7 percent, petroleum 2 percent, and miscellaneous renewables 2 percent.[6] From these statistics it's obvious that at 2 percent, oil is no longer a major factor in electrical generation in the United States. It's still important, but not crucial. That's the good news. Now the bad news. Natural gas unfortunately now represents about 18 percent of the total. Natural gas supplies in the United States have been stretched to the limit to meet demand in recent years, and that situation is about to get worse. A lot worse.

Beginning around 1999, almost 95 percent of new electrical generating plants built in the United States have been fueled by natural gas. Why the utilities decided to replace one finite fossil fuel with another is a legitimate question. And why they decided to do it *now* is an even larger question.

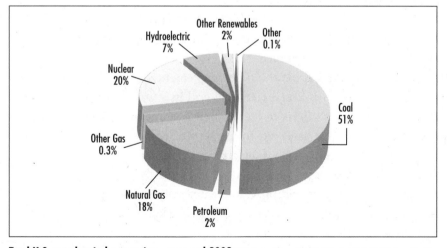

Total U.S. net electrical generation—year end 2002. U.S. Energy Information Administration (EIA), Electric Power Monthly, January 2004.

The idea was to replace dirtier, coal-fired generation with a cleaner fuel; natural gas emits about half of the CO_2 produced by coal when combusted. But in retrospect, this was a terrible mistake, as the timing could not have been worse. By the end of 2003, more than three hundred new gas-fired power plants had been constructed in the United States with a collective price tag in excess of $100 billion. Unfortunately, this took place at virtually the same time that U.S. natural gas production appears to have peaked, according to the Energy Information Agency.[7] Since a large, modern, gas-fired electrical generating station with a 500-megawatt capacity (one megawatt equals one million watts) can consume as much as 80 million cubic feet of gas a day, these hundreds of new plants will put a severe strain on dwindling gas supplies.[8] That, in turn, will put the reliability of the national electrical grid in serious jeopardy in the very near future (more on natural gas later in this chapter).

But there are more problems with the U.S. electrical grid. Except for hydropower (slightly less than 7 percent) and various renewables (2 percent), the remaining 91 percent of generation is based on finite fossil fuels and (extremely problematic) nuclear energy. As we enter the combined peak oil and natural gas calamity, there will be enormous political pressures to expand the use of "clean coal" (an oxymoron if there ever was one) and nuclear energy to maintain the struggling national electrical grid. The basic argument is going to be, "Do you want to have the lights on, or do you want them off?" Thanks to the adamant resistance to renewables by the coal, oil, natural gas, and nuclear industries over many years, we now face terrible and rapidly diminishing choices.

One positive change that should be implemented immediately, however, is to restructure the national electrical grid in a way that would allow for the integration of all types of local renewable energy generation. Locally generated power also saves up to 69 percent of the energy lost through long-distance transmission of electricity from large, centralized power plants. This generation could be either individually or community owned and locally based; it would provide greater stability of supply throughout the grid, and is the best protection against blackouts and possible terrorist attacks. Small steps in this direction have already been taken, but most large utility companies resist this commonsense strategy strenuously because it threatens their market dominance.

Space Heating

Space heating is the eight-hundred-pound gorilla in the room that few people seem to be paying much attention to. That's about to change dramatically. One-quarter of all energy in the United States is consumed in heating buildings. Billions of Btus are expended every heating season. Unfortunately, the vast majority of this energy still comes from nonrenewable resources such as coal, oil, and natural gas. In fact, natural gas is used to heat 55 percent of all U.S. homes. Combine that with 5 percent for liquefied petroleum gas (LPG), and gas now accounts for 60 percent of home heat. Electricity accounts for another 29 percent, while heating oil is a relatively modest 7 percent. Wood heats another 2 percent, while the remaining 2 percent is composed of kerosene, coal, solar, and so on. The use of fossil fuels and electricity (which, as I mentioned earlier, is generated mostly by burning fossil fuels) accounts for about 95 percent of home heating in the United States. That's downright scary. Heating is unquestionably the single largest energy expense in most homes, accounting for as much as two-thirds of annual energy bills in colder parts of the country. And in those regions, winter heat is literally a matter of life and death. Without heat, most homes and larger buildings can quickly become uninhabitable, especially if water pipes freeze and burst.

Ironically, the crunch in home heating is not going to be triggered primarily by high oil prices (though people who heat their homes with number 2 heating oil will pay dearly for their winter heat in the years ahead). The real crisis will come when the dwindling supply of natural gas can no longer meet demand and prices skyrocket. The 60 percent of Americans who heat their homes directly with gas will be left wondering what hit them. And the other 29 percent who use electricity will get a real shock when the hundreds of new natural gas–fired electrical power plants begin to run out of prohibitively expensive natural gas. What are these people going to do? (For starters, they might want to read my 2003 book, *Natural Home Heating: The Complete Guide to Renewable Energy Options.*)

OUR ENERGY OPTIONS

Now that we have a better understanding of the big picture, let's take a look at our present energy options and their relative advantages and disadvantages.

While we have a large number to choose from, many are problematic, and in the long run, unacceptable. We'll start with the nonrenewables because they play such a major role in our current economy.

Oil

In the not too distant future, some people are probably going to look back on the late twentieth and early twenty-first centuries and ask why all that valuable oil was being burned for fuel when it should have been conserved for the production of plastics, pharmaceuticals, and other valuable high-tech products—like renewable energy systems. Questions like that may be hard to answer.

It is increasingly likely that the more pessimistic predictions about the arrival of peak oil are correct, and that human society is about to begin its long, painful descent down the back side of Hubbert's Peak sometime in the next few years. Consequently, the need to implement serious alternatives to oil couldn't be more urgent. But since virtually every alternative energy strategy requires vast amounts of oil-based inputs or processes to produce or deploy, it's increasingly obvious that whatever oil we have left should be carefully conserved to help us implement these renewable energy initiatives. But don't hold your breath. In the Last One Standing scenario that many governments seem intent on following, our remaining resources will probably be squandered on warfare and other desperate attempts to commandeer the remaining supplies of oil (and natural gas) around the world. The lion's share of that remaining oil and gas is located in the Middle East, and the battle for that oil has unquestionably begun.

But what other oil options do we have? A large amount of potential oil is presently locked up in *tar sands* and *oil shales*. There has been a lot of attention focused lately on the Canadian tar sands of northeastern Alberta as well as the oil shales of the American West. Neither of these resources shows much potential, however, despite repeated efforts to recover the energy from them. The tar sands (also referred to as oil sands) are not conventional oil at all, but rather a mixture of tar (or bitumen), sand, and clay. This unpromising mixture is very difficult and expensive to extract, separate, and process into a synthetic crude oil. The open-pit, strip mining process normally employed in its extraction is extremely energy intensive, uses vast amounts of water, and is environmentally devastating. Although the deposits are potentially huge—the equivalent of perhaps between one and two *trillion* barrels of oil—the fact that this oil is in the form of solid bitumen makes its recovery extremely problematic. One of the many difficulties with this process is that it depends on huge amounts of energy, especially natural gas—the same natural gas that many U.S. and Canadian energy planners are counting on to avert a gas shortage catastrophe in North America.[9] Do we want the natural gas, or the oil from the tar sands? We can't have both.

The enormous reserves of oil shale in the western United States suffer from similar problems. The "oil" from oil shale is actually *kerogen*, a waxy hydrocarbon that is a precursor to petroleum, contained in a hard sandstone called marl. A ton of this rock yields about a barrel of oil. Oil shale has been described by some of its numerous critics as "the energy source of the future—and always will be." That's because the energy needed for recovery will probably exceed the energy represented by the resulting oil, making the whole enterprise a net energy loser. All previous attempts to recover this oil economically have failed. Exxon closed its $5 billion Colony II project in western Colorado in 1982. Other oil companies abandoned their projects the following year.[10] The last company, Unocal, finally gave up in 1991. The traditional method of extraction was via strip mining, which is environmentally destructive. Recent experiments at heating the deposits in situ by drilling wells and then electrically heating the kerogen over time and converting it into oil and gas have shown some promise. Major questions remain, however, regarding whether this can be accomplished economically, safely, and on a commercial scale. But that hasn't deterred the Bush administration from promoting this questionable strategy—as long as it is at taxpayer expense. A series of deadlines in the new energy bill will force the U.S. Bureau of Land Management to lease lands in Colorado, Utah, and Wyoming for oil shale and tar sands development by late 2007.

Synthetic Oil

During World War II, Germany produced a lot of synthetic oil from coal because the country had almost no oil of its own. The method of conversion, known as the Fischer-Tropsch process, was developed in the 1920s by Franz Fischer and Hans Tropsch. Although Germany abandoned this strategy at the end of the war, the Sasol Company of South Africa uses a proprietary version of this process to produce a variety of synthetic petroleum products, especially diesel fuel, from the country's vast coal deposits. Sasol produces more than 150,000 barrels of high-quality fuel daily. This strategy was initiated partly because of South Africa's isolation during the Apartheid era. Although it's a viable, proven technology, Fischer-Tropsch is hampered by high capital, operation, and maintenance costs. Nevertheless, a number of oil companies are pursuing pilot projects in several countries and some observers predict that Fischer-Tropsch fuels will be used in high-end diesel fuel blends.[11] Unfortunately, this process still relies mainly on nonrenewable, fossil fuel inputs (more on a renewable version of Fischer-Tropsch later).

Natural Gas

In addition to the crucial role natural gas plays in the space-heating and electrical-generating markets, it is also used to produce vital nitrogen fertilizer for agriculture, as well as vast amounts of plastics, fabrics, packaging, and other products. Unfortunately, the United States faces a looming natural gas catastrophe similar to peak oil. Despite massive exploration and drilling efforts, supplies of natural gas in the United States started to decline in 2001. Similar declines are now occurring in Canada and Mexico, the main sources of imports for the U.S. market. But until fairly recently, North American natural gas producers refused to admit that they had a problem with supply. This changed abruptly in 2003—amid turmoil in the fertilizer industry and rising concerns in the chemical and power industries—with the admission by gas producers that supply was, indeed, apparently in decline.[12] This sudden disclosure has caught many consumers and businesses by surprise, since natural gas has been touted for decades as a cheap, clean, and almost inexhaustible fuel source. While "clean" is debatable, natural gas is definitely no longer cheap or inexhaustible. The United States and Canada (and, to a lesser extent, Mexico) are now entering a serious natural gas crisis, from which there is no quick or easy escape.

The main response by the Bush administration has been to push for relaxed air pollution standards to allow for greater coal generation of electricity, and to encourage exploration and drilling for natural gas in more environmentally sensitive areas, as well as greater reliance on unconventional gas such as *coalbed methane* (CBM) as well as so-called tight sands and gas shale production. One of the main problems with CBM is the large quantity of wastewater that is produced, which contains many contaminants that increase its salinity to high levels. This wastewater can kill existing vegetation and has a toxic effect on crop and range lands. *Tight sands gas*, which is held in reservoirs where flow is considerably restricted, requires much more intensive drilling. This is bad news for the residents of the Rocky Mountains region of the United States where more than half of this gas is located.[13]

There is another response to the looming natural gas crisis that is supported by the Bush administration and the industry—the importation of vast amounts of *liquefied natural gas* (LNG) from overseas to make up for the growing domestic production deficit. Since LNG occupies only a fraction of the volume of natural gas, it is prepared for shipment at the country of origin by supercooling and then regasifying it at the import terminal. This strategy is fraught with multiple dangers. One problem is that LNG requires a complicated and extremely expensive ($3 to $10 billion) infrastructure, called an LNG train, composed of pipelines, terminals, double-hulled cryogenic tankers, and regasification terminals at the receiving end. Due to the extremely cold temperatures required (–260°F), this process is also energy intensive, and about 15 percent of an average shipment of LNG can be consumed in its transport from, say, Qatar to the east coast of the United States, including refrigeration and the energy needed for regasification. The total cost of the necessary LNG infrastructure for a large-scale import program relying on numerous LNG trains might be in the $200-billion range, which could seriously test America's financial resources.[14] Even after spending all this money, the strategy is ultimately flawed because global natural gas is expected to peak five to fifteen years after oil peaks.

Then there are the dangers of accidental or deliberate explosions. There have been a number of serious LNG accidents, especially the January 19, 2004, explosion in Skikda, Algeria, that destroyed a large amount of infrastructure and resulted in the deaths of twenty-four people, with seventy-four injured and seven others missing and presumed dead.[15] Although there is a good deal of disagreement about what would happen if a modern LNG tanker or facility were hit with a rocket or suffered a serious accident, this scenario has ignited a firestorm of protest to the siting of LNG terminals in the United States. No one wants to have the equivalent of a nuclear bomb in their backyard. In addition, since it takes between three and seven years to move a LNG terminal from planning stages to completion, it is likely that these facilities will not arrive in time to avert the gas shortage they are intended to avoid. Finally, the major security risks involved with relying on yet another imported source of finite fossil fuel from nations that don't like the United States very much should be enough to give pause to even the most ardent supporter of LNG.

Coal

Coal is still relatively abundant, especially in the United States, which has the world's largest reserves. On the downside, burning coal produces more carbon dioxide and particulate pollution than burning oil or gas. It also produces sulfur dioxide, a gas that contributes to acid rain, and mercury, which poisons the environment and people. Underground coal mining is difficult and dangerous, and both underground mining and surface strip mining are environmentally ruinous. But in addition to being the dirtiest of fossil fuels and a major contributor to global warming, coal suffers from a dilemma similar to oil: the highest-quality and easiest-to-recover reserves have already been mined. As a result, the energy-to-profit ratio for coal has been declining in recent decades. While there is still plenty of coal, the remaining recoverable reserves will eventually be exhausted at increasing financial cost, while we simultaneously devastate the environment.

Despite industry claims for "clean coal technology," there simply is no such thing. Coal is a dirty and polluting, nonrenewable fuel. The "clean coal" claims refer only to the smokestack emissions of coal-fired electrical generating plants—totally ignoring the environmental devastation caused by mining. Just ask the folks in states like West Virginia about the destruction that mining is wreaking on their communities. Mountaintop-removal miners have leveled more than three hundred thousand acres of the state's rolling mountains by blasting them away or bulldozing them into adjacent valleys. The environmental and social consequences have been staggering. Nevertheless, the coal industry and its supporters in Washington, D.C., are gearing up for a major campaign to increase coal use in the years to come, at the very time when we need to be reducing our dependence on coal. With coal prices up and coal-fired power plants cited by the Bush administration as the energy source of the future, strip mining in Appalachia and elsewhere is increasing at a frightening pace.[16]

In early 2003, the U.S. Department of Energy announced a presidential initiative to build "FutureGen," a $1 billion demonstration project that in ten years would lead to the world's first "emission-free" plant to produce electricity and hydrogen from coal while capturing greenhouse gases.[17] Electricity and hydrogen from coal: you don't have to be clairvoyant to see where this idea is heading—more reliance on coal. Confirming this unfortunate trend, at least 114 new coal-burning power plants are currently in the building or permitting stages around the country. According to a 2006 report from the U.S. Energy Information Administration, U.S. power consumption from coal is expected to rise 1.9 percent per year through 2030.[18] Clearly greenhouse gas emissions are not about to drop in the United States, immediately or in the foreseeable future, without a dramatic change in national energy policy.

Nuclear

Once viewed by its supporters as a limitless source of cheap power, nuclear energy is now viewed by most people as being neither. It has proven to be far too dangerous and expensive, and the nagging issue of what to do with large quantities of highly radioactive waste has yet to be resolved. The Yucca Mountain nuclear waste dump in Nevada, which was supposed to solve the storage problem for these long-lived toxic wastes, is located in the third most seismically active area in the state. The dump is being constructed over the strenuous objections of local residents, and it is not clear whether it will ever be used for its intended purpose. Meanwhile, nuclear waste continues to accumulate at nuclear power plants around the country, overwhelming the containment pools that were originally built as "temporary storage" for this material. On-site spent-fuel storage casks have been developed as another "temporary" fix for the overflowing containment pools. How many more "temporary" strategies will it take before the industry finally drowns in its own poisonous garbage?

What's more, even after fifty years of electrical generation, nuclear fission electrical-generating plants are still not economical. These facilities require huge subsidies to be constructed, and these costs are never figured into the cost of the electricity generated by them. If the industry had to actually cover all of its own costs without relying on taxpayer subsidies and bailouts, it would never survive in the open market. Aside from the strong public opposition to nuclear power since the Three Mile Island accident in

1979 and the 1986 Chernobyl disaster in the former Soviet Union, the enormous financial burden associated with plant construction is one of the main reasons why private companies in the United States have had little enthusiasm for building new nuclear power stations for the past thirty years or so.

An alternative nuclear strategy, breeder reactors, has been touted for many years as a way to extend the potential for nuclear generation of electricity well into the future. Unfortunately, these reactors require the use of plutonium-239, an extremely persistent radioactive poison, which can also be used for nuclear-bomb making. In a post-peak-oil environment where a lack of resources, and, especially, a lack of adequate security are very real possibilities, this strategy would not seem to make much sense. But there have been many other problems associated with this program, and the only prototype breeder reactor in the United States has been shut down, while similar programs in the United Kingdom and France have also been terminated.

But there are other difficulties with nuclear power. Nuclear power needs uranium to produce the fuel that powers the generators. There is about a fifty- to one-hundred-year supply (depending on whose figures you use) currently available to fuel the 400 existing nuclear power plants worldwide (109 in the United States). However, if a new, large-scale new-generation nuclear power program were instituted in the near future—which appears increasingly likely in the United States—that supply of uranium would be exhausted much sooner. But perhaps the biggest problems of all are that the costs of constructing these new plants may simply be too high in a post-peak-oil environment where the cost of everything will skyrocket along with the price of oil, and that the time frame for completing the projects simply extends too far into the future to avert the coming energy crisis.

The Holy Grail of the nuclear industry is what is known as *nuclear fusion*, which mimics the thermonuclear process that powers the sun. The potential power output from a fusion reactor is enormous, but the challenges to its practical success are equally huge. The closest we have come to replicating this process so far is the creation of the hydrogen bomb, and containing and harnessing that power is a daunting problem. Nevertheless, in June 2005 France was selected as the site for a €10-billion (euros) nuclear fusion reactor. The International Thermonuclear Experimental Reactor (ITER) is a collaboration among the European Union, the United States,

Russia, Japan, South Korea, and China, and it represents the most expensive joint scientific project after the International Space Station.[19] While supporters have hailed the ten-year initiative as a major first step, critics complain that the project is a colossal waste of money that could have been better used for the implementation of currently available renewable energy technologies such as wind farms. But the main problem with ITER is that it is only a scientific demonstration project, not a commercial electrical-generation facility. Even under the most optimistic scenario (which is highly unlikely) a functional commercial plant could not reasonably be expected before 2050. We simply don't have forty-plus years to wait for some magic technological nuclear breakthrough, which brings us back to wishful thinking—again.

Despite all of these problems, the Bush administration is firmly committed to reviving the nuclear industry. "Of all our nation's energy sources, only nuclear power plants can generate massive amounts of electricity without emitting an ounce of air pollution or greenhouse gases," President George W. Bush said as he signed the energy bill on August 8, 2005, while glibly ignoring the whole nuclear waste issue. "And thanks to the advances in science and technology, nuclear plants are far safer than ever before. Yet America has not ordered a nuclear plant since the 1970s. . . . With the practical steps in this bill, America is moving closer to a vital national goal. We will start building nuclear power plants again by the end of this decade."[20]

The nuclear industry has finally gotten what it has been waiting for all these years. This should come as no great surprise, since the industry helped to write the energy bill. But relying on nuclear energy to save us from the impending energy crunch is a devil's bargain if there ever was one. While we may keep the lights on for a while, we risk polluting the planet with life-threatening toxic wastes that may eventually escape into the surrounding environment and could haunt our descendants for centuries. And the specter of nuclear materials "escaping" into the hands of rogue nations or organizations is even worse.

Before we move on to renewables, a few brief observations are in order. There is one characteristic that virtually all of the nonrenewable energy strategies we've looked at so far have in common—they are large-scale, corporate-dominated industries that are not easily downsized to local scale for individual or community use. And they have the unfailing support of the administration and many members of Congress who represent these corporate interests, rather than the American people. Under these unfortunate circumstances, it is unrealistic to expect much meaningful help from the federal government, which is essentially paralyzed by these powerful special interests. (One can always hope for a more enlightened behavior, but I'm not holding my breath.) It should also be fairly clear by now that most of these nonrenewable strategies are part of the problem, rather than part of the solution—despite the constant spin to the contrary by their lobbyists and supporters inside the Beltway. Now it's time to take a look at the many renewable energy options we have to work with.

RENEWABLES

The remainder of this book will be focused on renewable energy strategies. While these strategies are not perfect, they are the best hope we have to help create the new decentralized, localized economy that we will be forced to adopt after we pass over the top of Hubbert's Peak for oil and natural gas. Actually, the planning for this new economy has been going on for many years at the local level. And some of these initiatives, while generally small in scale, offer considerable promise for the people who are involved in them. They also offer exciting, positive models that could be replicated in other locations by similar groups of dedicated people who are willing to work together for the common good of their communities. And that is a key component of these examples: building community while we build a new society from the ashes of the old, hyperconsumptive society we are about to leave behind.

The alternative to this—a variation on the Last One Standing strategy—involves hoarding, competition, fear, paranoia, and violence, and is associated with a go-it-alone, survivalist mentality. But that's not what I am advocating. While survival is also a goal of the cooperative model I'm suggesting, there is strength in numbers, and a cohesive, cooperative community is far more likely to make it through the difficult times ahead than a lone cowboy.

Before we look at the individual strategies, I want to make a few general observations about renewables. First, it's important to put renewables into their proper perspective. If our total energy use in the United States was represented by a pie, only a thin slice—roughly 6 percent—would represent renewables. And most of that tiny slice is represented by biomass energy (47 percent) and hydroelectric power (45 percent) for space heating and electrical generation. The remainder of the slice represents geothermal energy (5 percent), wind (2 percent), and solar energy (1 percent) of this tiny slice of the pie (*not* 1 percent of the larger pie). Despite the huge increases in both wind and solar installations in recent years, taken together, they still only represent a miniscule 0.18 percent of our total energy picture. This should help you understand the immense challenge we face in this country, and why I am proposing these strategies on a small-scale, local level rather than as large-scale national solutions. While large-scale renewable initiatives are fine, and we should encourage their development where appropriate, there simply isn't the time or the money to expand renewables to the point where they will solve the present national energy crisis. But there is the possibility that we can still offer some relief at the local level by developing these renewable resources immediately, with the hope that this will buy us enough time for the rest of the nation to eventually catch up.

Also, I do not blindly advocate the use of renewable energy strategies unless they are less polluting or destructive to the environment than their conventional fuel alternatives. In some cases, that may be a hard call, but in general, the advantages of these renewable strategies far outweigh the

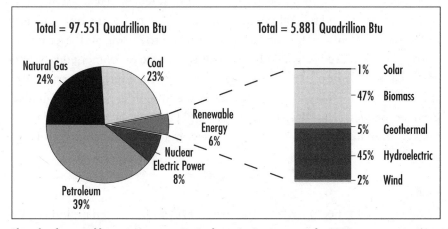

The role of renewable energy consumption in the nation's energy supply, 2002. EIA.

disadvantages, as long as they are implemented intelligently. As we will see, that's not always the case. Nevertheless, we will unquestionably be faced with some difficult decisions, and we must be prepared to make some compromises along the way. There are no perfect solutions, only imperfect solutions in an imperfect world. Having said that, here are our main options.

Solar

The sun is our planet's greatest source of energy. Every day, more solar energy strikes the earth than all of us consume in equivalent electric power in twenty-seven years, according to the National Renewable Energy Laboratory. We can use solar energy in two main ways, *passively* and *actively*. Passive solar energy is captured mainly through the design and orientation of a building, or through some of its elements, to take advantage of exposure to the sun. A well-designed passive solar home or building is unquestionably the best solar strategy you or your community can follow. Active solar systems, on the other hand, involve installing special equipment that uses energy from the sun to heat water or living spaces or to generate electricity. Active systems are a good retrofit strategy for existing homes and buildings, but generally require a lot of energy to manufacture.

Most active-solar systems fall into one of two categories: *photovoltaic* (PV) systems that produce electricity, and panels that produce heat for *solar domestic hot water* (SDHW). The cost of these systems has dropped substantially since they were introduced in the 1970s. In recent years, a number of municipal, state, and federal programs have been encouraging homeowners to add solar systems to their roofs. Depending on the siting and orientation of your home, there is a good chance that some type of solar energy system could be installed on or near to your home. Photovoltaic solar collectors that look like standard roofing shingles or metal roofing are now available from a number of manufacturers. Active and passive solar strategies can be used by individual homeowners, businesses, and communities in a number of different ways.

Wind

Wind energy has been the fastest-growing segment of the renewable energy market in recent years. Wind turbines (often incorrectly referred to as "windmills") have come a long way since they were first developed in 1891. Early wind turbines were capable of generating a few kilowatts, enough to power a house or two. Today they are hundreds of times larger and capable of powering entire communities. Many wind turbines being installed today generate 1.65 megawatts, and have blades equal to the wingspan of a Boeing 747. Even larger turbines, in the 3- to 4-megawatt range, are beginning to show up here and there; the largest currently operating has a generating capacity of 5 megawatts. Wind power does not produce air emissions, generate solid waste, or use water. Wind turbines do occasionally kill birds (though modern turbines are much less likely to do so), and some people object to their visual impact on the landscape. Others consider wind turbines to be beautiful, and describe the slowly spinning blades as "kinetic sculpture."

During and after the OPEC oil embargo, there was a flurry of wind-turbine activity in the United States, and some real advances were made in the technology. But when the Reagan administration pulled the plug on the energy tax credits and incentives that had encouraged the installation of renewable energy systems nationwide, the U.S. wind-power industry collapsed. Denmark, on the other hand, understood the incredible long-term potential for wind power, and after a twenty-year partnership between government and industry, emerged as the world leader in the field. Wind farms now generate 20 percent of Denmark's total electrical energy. There is a wide range of different-sized wind turbines for virtually every potential user, from the individual homeowner to community cooperatives, all the way up to the largest industrial or utility-scale installation.

Water

Water power (or hydropower, as it is often called) uses the energy of falling or moving water. In a hydroelectric power installation, the energy of falling water spins the turbines that generate the electricity. In a hydromechanical installation (rare these days) a turbine or waterwheel turns shafts that directly power various pieces of machinery through a series of gears or pulleys and belts. Hydropower does not produce any emissions or solid waste, but can cause other problems, depending on the size, design, and location of the dam and power plant. A large dam can disrupt fish migration in rivers, displace people and wildlife, and eventually cause silt buildup behind the dam, ending its useful life. Smaller "run of the river" hydropower installations that don't back up large amounts of water behind them have lower negative impacts.

Although hydropower offers a lot of advantages, most of the commercial-scale dam sites in the United States have already been developed, and it is unlikely that any more large dams will be built. While most of the large-scale hydro sites in the United States are already utilized, there are still numerous small-scale, local sites that could be developed (or redeveloped). About 5,677 sites in the United States, with a capacity of about 30,000 megawatts, have been identified by the U.S. Department of Energy.[21] If all these sites were developed, this could boost hydroelectric production by about 40 percent. The sites are mostly located on smaller rivers and streams, and offer considerable potential for local, small-scale initiatives.

An interesting subcategory of hydropower is ocean energy. Most ocean energy plants are small and experimental. But there is some interesting long-term potential for this renewable, generally nonpolluting energy source, and there has been renewed interest in ocean energy in the past few years. There are three main ocean technologies: wave energy, tidal energy, and ocean thermal-energy conversion. A number of promising pilot projects have been successful in harnessing some of these resources recently, and a few commercial-scale initiatives are about to be launched.

Biomass

Sunlight makes plants grow, and the organic matter that makes up these plants is known as *biomass*.[22] The oldest use of biomass for energy is wood heat. Biomass can also be used to generate electricity, and to make solid fuels such as wood chips or pellets for heating. (Biomass can also be used to create liquid biofuels, which I will describe in a moment.)

Biomass electrical energy (also referred to as biopower) can come from crops grown especially for energy production, such as switchgrass or trees, or it can be generated from organic wastes, such as methane gas from landfills or methane digesters on farms. There are four main types of biopower: direct fired; cofiring; gasification; and small, modular systems. Most biopower plants use the direct-fired method. They burn the feedstock directly to produce steam, which spins a turbine that generates electricity. Cofiring is a strategy used mainly by coal-fired electric generation plants to reduce emissions, especially sulfur dioxide, by mixing coal and bioenergy feedstocks.

Gasification systems convert biomass into gas (sometimes called *syngas*), which is then burned in a gas turbine or internal combustion engine that spins an electric generator. Methane gas from landfills or methane digesters on farms can be also be used to generate electricity. Small, modular systems can be fueled by any of these methods and are used in small-scale community or individual installations. The farm-based and community biomass systems, in particular, offer a lot of potential for local, cooperative initiatives. In recent years, biomass has only accounted for about 3 percent of energy used in the United States, but that's about to change dramatically.

Liquid Biofuels

Biomass can also be used to create liquid biofuels such as ethanol, biodiesel, and methanol. Ethanol and biodiesel, in particular, have been receiving quite a lot of attention lately. Ethanol is made by converting the carbohydrate portion of biomass into sugar, which is then converted into ethanol in a fermentation process that is similar to brewing beer. Ethanol is the most widely used biofuel today with current U.S. production of 4.8 billion gallons annually, based mainly on corn. However, ethanol produced from cellulosic biomass feedstocks (rather than corn) is currently the subject of intensive research, development, and demonstration projects and offers a lot of promise.[23]

Although ethanol has been around for many years, biodiesel is a relative newcomer, especially in the United States, where it has mainly been used and promoted by a small group of backyard enthusiasts. But not anymore. Biodiesel has emerged in the past few years as a mainstream vehicle and home-heating fuel. U.S. biodiesel production has vaulted from 500,000 gallons in 1999 to 75 million gallons in 2005, and is expected to double to at least 150 million gallons by the end of 2006. The industry is expected to continue to expand exponentially to meet the rapidly growing demand.

Biodiesel can be easily made through a simple chemical process from virtually any vegetable oil, including (but not limited to) soy, corn, rapeseed (canola), cottonseed, peanut, sunflower, mustard seed, and hemp. But biodiesel can also be made from recycled cooking oil or animal fats. There have even been some promising experiments with the use of algae as a biodiesel feedstock. And the process is so simple, biodiesel can be made by virtually anyone, although the chemicals required (usually lye and methanol) are hazardous and need to be handled with extreme caution. Both ethanol and biodiesel offer considerable opportunities for small-scale, local, cooperative production initiatives, especially at the small family-farm level.

Lastly, biomass can also be gasified to produce a synthesis gas composed primarily of hydrogen and carbon monoxide, also called wood gas, biosyngas, syngas, or producer gas. The hydrogen can be recovered from this syngas, or it can be catalytically converted to methanol. It can also be converted using a Fischer-Tropsch catalyst into a liquid with properties similar to diesel fuel, known as Fischer-Tropsch diesel. Since it is based on biomass rather than coal or natural gas, this variant on the Fischer-Tropsch process is renewable. This is an area of intensive research and development work at the present time, and commercialization is expected in the next few years. Because liquid biofuels are so important in our energy future, and because there is so much activity to cover, they get separate treatment in chapter 6.

Geothermal

Geothermal uses heat from the interior of the earth to produce clean, more-or-less sustainable energy. In some locations, naturally occurring steam, superheated groundwater, or a heat-transfer fluid are used to generate electricity. About 2,800 megawatts of geothermal electricity is generated annually in the United States, mostly in California and to a lesser extent in Nevada, Hawaii, and Utah. Some geothermal resources are also utilized for what is referred to as "district heating" of buildings and sometimes for agricultural and industrial processes. Although geothermal sources can provide heat for many decades (up to several hundred years in some cases), most are eventually depleted, so they are not strictly renewable like wind farms or solar installations. Because of the limited number of suitable geothermal sites, this is admittedly not a strategy for everyone.

But you don't need to live in a geologically active area to tap into ground heat. That's because, in most locations, within the top fifteen feet or so of the earth's surface, ground temperatures remain fairly constant year-round at between 45° and 60°F (7° to 16°C). Geothermal exchange heat pumps can tap into this resource and heat or cool a building. In the winter, heat is extracted from fluid in pipes buried in the ground (or sometimes from pond or well water) and then increased with the help of a compressor. The heat is then distributed through the home or building, usually by a series of ducts. In the summer, the process is reversed, and the heat from your home is conducted out of the house and dissipated into the ground. Geothermal exchange is the most energy efficient, environmentally clean, and cost-effective space-conditioning system available, according to the Environmental Protection Agency.

Hydrogen

Hydrogen has some excellent qualities. It can be combusted like other ordinary fuels, or it can be converted to electricity in fuel cells, forming water or steam as an exhaust product, with virtually no negative environmental impact. And there's no problem with supply: hydrogen is the most abundant element in the universe. Consequently, the "hydrogen economy" has been promoted as the wave of the future and the answer to our global energy woes. But I have my doubts. So do an increasing number of other observers.

Strictly speaking, hydrogen is not really a fuel source, but rather a renewable energy storage and transport medium that must first be produced from some other source. One method is to extract it from water in a process known as *electrolysis*. And that's a problem, because the electrical energy required for this process under existing technology is substantial, and would almost certainly require the construction of hundreds of new nuclear power stations to supply the electricity. Hydrogen can also be separated from water by superheating, but this strategy (like many associated with hydrogen) is a net energy loser. Hydrogen currently takes about twice the amount of energy as other fuels to do same amount of work. Hydrogen is also bulky to store and transport and virtually none of the massive production and distribution infrastructure that would be required is currently in place. This new infrastructure would probably cost billions of dollars, which seems increasingly unaffordable in a post-peak-oil economic scenario.

But fuel cells themselves present additional problems. While fuel cells can operate at efficiencies two to three times greater than an internal combustion engine and have no moving parts, the primary fuel today is natural gas or methanol. With these fuels, a fuel cell produces carbon dioxide, just like a conventional internal combustion engine, and, as noted earlier, supplies of

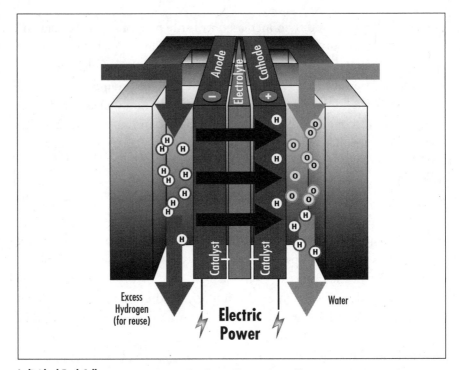

Individual Fuel Cell. U.S. Department of Energy, Office of Energy Efficiency and Renewable Energy.

natural gas are already running short in North America. This hasn't kept the natural gas industry from enthusiastically promoting the new hydrogen economy. It's also increasingly clear that the "hydrogen economy" has been hijacked by the coal, natural gas, and nuclear industries, who see it as a way to ensure their continued market dominance. Many observers now view the hydrogen economy as simply a Trojan horse for these dinosaur industries. This also goes a long way toward explaining the Bush administration's enthusiastic support of hydrogen.

James Howard Kunstler sums up the hydrogen question this way: "What all this boils down to is that a hydrogen-powered automobile system and all its supporting infrastructure cannot be substituted for an oil-based system under any plausible equation currently understood. Without ubiquitous fuel stations and methods for supplying them that are both economically and logistically rational, there is no basis for such a system."[24]

Even James Woolsey, former CIA director under President Clinton, has serious doubts about hydrogen. "In recent decades Americans have benefited from cheap oil, thinking that someday, the hydrogen economy would save us," Woolsey said in a speech at a renewable energy conference in Las Vegas, Nevada, in April 2006. Woolsey called for other renewable energy programs that do not require extensive research and development, rather than research into hydrogen as a fuel.[25] I agree. While there may be some uses for hydrogen in certain sectors of the economy in the years ahead, it is unlikely to become the silver-bullet technology that most people are expecting (or hoping) it to be. Maybe I'm wrong, but I think it's important to be realistic about what our real options are in the short term and not waste our time and money on impractical fantasies.

OTHER STRATEGIES

There are a number of other energy strategies that have received a lot of attention in recent years, but I'm not so sure about them either. *Methane hydrates* is a prime example. Methane hydrates (also known as methane ice) occur when ice crystals form in a cage-like pattern that traps methane within. An enormous quantity of methane hydrates is thought to be trapped in the ocean sediments at low temperatures and extreme pressure below about a thousand feet, and under permafrost in Siberia and other land near the poles. There is a great deal of disagreement, however, about methane hydrates in general, and a number of serious problems regarding its possible recovery. One concern (by now a familiar refrain) is that it probably will take more energy to recover the methane hydrates than they will yield, meaning that this exercise would simply be uneconomical. But methane ice is also hazardous, and attempts to recover it have led to explosions and damage to drilling equipment. And since methane hydrates tend to be extremely unstable when disturbed, uncontrolled accidental escape of methane into the atmosphere is almost assured. Methane is more than twenty times more harmful as a greenhouse gas than carbon dioxide, so releasing more of it would greatly exacerbate global warming.[26] But the main problem is that methane hydrates are a finite resource that would eventually be exhausted (if they can ever be recovered safely and economically).

Thermal depolymerization (TDP) is another possible energy source. This involves claims about turning virtually any carbon-based waste material, including "turkey offal, tires, plastic bottles, old computers, municipal garbage, corn stalks, paper-pulp effluent, infectious medical waste, oil refinery residues, even biological weapons such as anthrax spores" into usable hydrocarbon fuels, gas, and valuable minerals. While there is a full-sized TDP demonstration facility in operation in Missouri, to a large extent it depends on the waste from a nearby commercial-scale turkey processing plant for its primary feedstock. But turning turkey guts into oil is hardly high-tech alchemy (try skimming the turkey fat out of the broiler pan next Thanksgiving, and you get the general idea). What's more, if the large-scale agribusiness system collapses sometime after peak oil and peak natural gas (as is likely), there goes the easy feedstock. The same can be said for most other potential feedstocks, which are also based on a wasteful cheap-oil economy.

Although the media has breathlessly reported the TDP story without much scrutiny, there are serious critics of the whole concept. "Now along comes one more scheme for playing on the gullibility of a public that . . . actually believes that a complex technological society, such as ours, can rip and strip the earth of all its resources, use them transiently, then somehow destroy them all, and still continue to leave a thriving planet for future generations. As though the earth is some kind of a magic lamp we can rub and the genie will continue to bestow upon us any gift we request. This concept is idiotic, and any company that seeks to effectuate the 'getting rid of' part of this scheme is selling a bill of goods leading to planetary suicide." So says Dr. Paul Palmer, author of the 2005 book *Getting to Zero Waste*.[27] It would seem that while TDP may have some interesting short-term potential, its long-term prospects are debatable.

Another questionable silver-bullet strategy is *zero point energy*. This involves harnessing the "dark matter" of the universe, and although there have been repeated claims that a breakthrough is "just around the corner," I'm not about to invest in this scheme any time soon. My main concern is that while we wait around for someone to harness the "dark matter" of the universe, we will end up in the dark because we weren't focusing our attention on keeping the electrical grid up and running with available technology. Okay, show me, and then I'll be an enthusiastic supporter. In the meantime, we need carefully to choose practical, proven technologies that we can rely on in an economy that is going to be increasingly hard-pressed to provide the basic goods and services we need for survival. We cannot afford to "wait for the magic elixir."

CONSERVATION

Last, but by no means least, there is conservation. This is both a practical, proven, and immediately usable strategy. Conservation is also the least expensive strategy we have available and the least harmful to the environment. It can be used to reduce our consumption of a wide range of resources, including, but not limited to, fossil fuels, electricity, and water. Conservation also reduces the need to build new electrical-generation plants and other sources of energy. These reductions can be achieved through a combination of initiatives, including the use of energy-efficient building materials, appliances, and other technologies, coupled with imaginative thinking about living better with less. Best of all, conservation can be implemented at the national, state, community, or individual level. You don't have to wait for everybody else; you can get started immediately.

Here is one example. My home state of Vermont has established an award-winning program that conserves energy, helps businesses, and protects the environment all at the same time. Efficiency Vermont, a statewide "energy-efficiency utility," provides a comprehensive set of energy-efficiency programs to Vermont electric consumers. By combining what was formerly a loose patchwork of energy-efficiency programs throughout the state, Efficiency Vermont successfully achieves economies of scale and significant cost benefits. The program has also achieved twice the national average in energy savings of other states' efficiency programs.

Although the Efficiency Vermont program is unique, it could be replicated almost anywhere. Already, the states of Maine, Wisconsin, and Oregon are moving toward a state-run energy-efficiency model, and in January 2006 Hawaii launched a comprehensive integrated approach to reducing oil dependence based, in part, on the Efficiency Vermont model. This idea has also attracted interest among utility, governmental, and NGO representatives in Brazil, Australia, and China. By relying on the two simple principles of integration and conservation, Efficiency Vermont has achieved some remarkable energy savings and resource benefits for the

AMORY LOVINS: CONSERVATION PIONEER

The concept of conservation is not new. In the 1970s, many energy experts were predicting large increases in energy demand that would require the construction of numerous power plants to keep pace with demand. But Amory Lovins (who cofounded the Rocky Mountain Institute) proposed an alternate strategy—which he called the "soft path"—that relied on energy-efficient technologies and conservation to meet the demand. As part of this strategy, Lovins developed the concept of "negawatts," which allowed utilities and governments to compare the cost of conservation measures against the cost of increasing power production. Negawatts represent power saved from one use that is made available for another use. For example, a compact fluorescent light bulb uses about one-fourth as much energy as a standard incandescent bulb to produce a similar amount of light. Replacing one 100-watt bulb with one 25-watt compact fluorescent therefore "generates" 75 negawatts of saved energy that can be used somewhere else. The concept of negawatts eventually evolved into "negagallons" to offer city utilities a new way of evaluating water supply and demand. Using negagallons, Lovins was able to demonstrate that it was more economical to institute a wide array of conservation measures in one city than to build a new dam. Lovins's energy-saving and conservation ideas proved to be extremely useful and have been widely adopted by municipalities, utilities, and businesses around the world.[28]

entire state. These benefits include substantial savings of electricity, fossil fuel, and water. This self-funded, business-savvy, and environmentally friendly program has a bright future in Vermont and elsewhere.[29]

But there are other ways to achieve a grassroots-based movement of conservation that extends well beyond energy-efficient lightbulbs. This involves transforming our own lifestyles into something more sustainable. This is not a new idea, but it's an important one. For many years, new lifestyle pioneers have been adopting one of two separate but related strategies: *downshifting* or *voluntary simplicity*. Both of these approaches contain varying degrees of consumption-reduction strategies. Involuntary conservation is coming soon anyway, so we might as well get some prac-tice now.

The many renewable energy strategies that we will be looking at in the next six chapters are like the jumbled pieces of an enormous and complex puzzle. We need to study each of them carefully before we try to fit them all together to create a picture that makes sense. We'll begin with a detailed look at solar energy strategies.

CHAPTER TWO

SOLAR ENERGY

In the fall of 2000, my wife and I bought a house in Weybridge, Vermont. Built in 1956, the modest, Cape-style home was neither especially energy efficient nor properly oriented for passive solar adaptation. But we eventually realized that we did have a south-facing garage roof that would provide a good location for several solar domestic hot-water (SDHW) collectors. Of all the possible active-solar retrofits, SDHW systems are the most cost-effective. When the state of Vermont offered a solar incentive program in 2003 that would pay for part of the $4,000 installation cost, we decided to have a SDHW system installed.

The components of the system include two 4-by-8-foot flat-panel solar collectors, a small 35-watt photovoltaic (PV) panel to power the system pump, an eighty-gallon electric hot-water heater (which provides both a hot-water storage tank as well as electric backup heat for cloudy days), a heat exchanger, and various controls, valves, fittings, and pipe insulation. Because we live in a cold northern climate, our system was an indirect, closed-loop design (more on design later) that contains an antifreeze solution in the collectors and piping to avoid the danger of freezing. When the sun heats the antifreeze in the collectors, the solar-powered system pump circulates the antifreeze through the pipes and into the basement, where it passes through a heat exchanger mounted on the exterior of the hot-water tank. The heat exchanger then transfers the heat to the water in the tank. The piping between the storage tank in the basement and the collectors on the garage roof is insulated to minimize heat loss. Ever since the system was completed in early 2004, it has operated dependably on sunny and even lightly overcast days, providing some of our hot water during the winter months and virtually all of our hot water during the summer. The system should pay for itself in about ten years, and in the meantime it's essentially operating for free. Since domestic hot water is usually the second largest utility expense for most homes (space heating is the largest), installing a SDHW system can save you hundreds of dollars a year and reduce your consumption of fossil fuels at the same time. We have also completed several other renewable-energy-related retrofits to our home that I'll describe later in the book. In addition, we plan eventually to install some PV solar electric panels on the remaining unused section of our garage roof or on a solar tracker[1] in our backyard or perhaps both. If, like us, you have an existing home that does not lend itself well to passive solar adaptation, you may be able to add some active solar retrofits for domestic hot water, solar home heating, or photovoltaic electrical generation. Anything that you can do to reduce your dependence on fossil fuels is a step in the right direction.

SOLAR ENERGY STRATEGIES

Solar energy is a sustainable resource that is available almost everywhere on the planet. In fact, virtually all renewable energy strategies are either directly or indirectly related to the sun. Enough solar energy falls on the earth in forty-five minutes to power the whole world for a year. This energy source is free and, unlike fossil fuels, it is immune to rising energy prices or supply shortages, at least for the next few billion years. Solar energy can be used in many different ways to provide heat, lighting, and electricity. There are many strategies to harvest this free energy, including passive solar building design, active solar systems, and photovoltaic installations. In many respects, passive solar is the simplest, so we'll start there.

Passive Solar Design

Arguably the most cost-effective solar strategy of all is passive solar design, especially when it comes to heating and cooling your home or other structures. A properly designed and oriented passive solar building will absorb solar heat in the winter and avoid it in the summer. Best of all, passive systems operate for free and generally do not require electrical or mechanical devices to function, although some form of backup heat is usually needed in colder climates. When it is included in the original design of your home, a passive-heating system doesn't materially add to construction expenses because you are simply making more intelligent use of the basic elements

Solar domestic hot water collectors on the roof of the author's garage. Greg Pahl.

that would generally be in your house anyway. Unfortunately, passive solar design is ignored in the vast majority of housing built in the United States—especially tract housing in suburbia. This means that we are wasting huge amounts of energy in this country simply on account of lazy building design, something that up until now has been subsidized by cheap energy in general, and cheap fossil fuels in particular. We can no longer afford this waste.

Passive solar design is not rocket science, but there are a few basic principles involved. A solar home's primary design factor is its relationship to the sun. But you can't just take any home in any location and expect automatically to harvest solar energy. That's because, first, you need a suitable site. Unfortunately, some sites, particularly those with a southern exposure obstructed by tall evergreen trees, buildings, hills, or mountains, just won't work. The best site for a passive solar home is one with unobstructed southern exposure between the hours of 9 A.M. and 3 P.M. coupled with protective slopes or large evergreen trees on the north side to block cold winter winds (in this case, the evergreen trees are an asset). But even if you

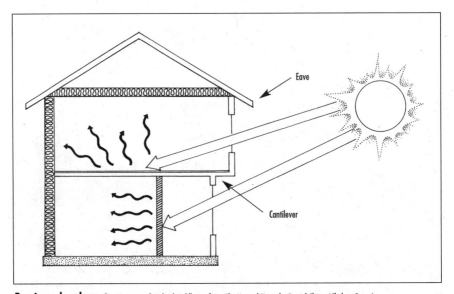

Passive solar elements. Courtesy of Michael Middleton, from *The Natural House* by Daniel Chiras (Chelsea Green).

have a good site, an improper alignment of your house can make an otherwise good plan fail. In general, a south-facing orientation that is within 30 degrees east or west of true south (the position of the sun at noon) will provide about 90 percent of the maximum solar collection potential. Southern orientation is crucial because the south side of a passive solar house is where all the action is from a solar energy perspective.

The other key element of good passive solar design is making sure that various parts of your house, especially the roof, are aligned with the sun's best angle, something that is determined by the latitude of your home's location and the seasons. This is helpful if you are planning on installing roof-mounted active solar collectors (more on active systems in a moment). The main thing to remember is that the sun is much lower in the sky during the winter and more nearly overhead in the summer in the Northern Hemisphere. This can work to your advantage in passive solar design, since during the winter the sun will shine in south-facing windows providing free solar heat, while in the summer it shines in less, or not at all if your roof overhang has been properly designed.

Every passive solar home has five main elements: the *aperture* (or collector), *absorber*, *thermal mass* (or heat sink), *distribution*, and *control*. Here are brief descriptions of these elements:

- The aperture is the large glass window area that allows the sun to enter the building.
- The absorber is the generally hard, dark, external surface of the heat-storage element.
- The heat-storage element or thermal mass is one of any number of possible materials (water, brick, concrete, adobe, rocks, and so on) located behind or below the absorber that store the heat produced by the sunlight.
- Distribution is the method used to circulate solar heat from the collection and storage locations to other areas of the house. A strictly passive solar house will use natural methods of heat transfer—conduction, convection, and radiation—while some designs also rely on mechanical devices such as fans and blowers (though avoiding mechanical devices whenever possible is best).
- Control of summer overheating is provided by roof and cantilever overhangs, awnings, blinds, or even trellises. Other (more active) controls might include thermostats for fans, vents, and other devices that assist or restrict heat flow.[2]

While the basic concept of passive solar is fairly simple, getting the actual design elements sized and arranged properly is not so easy. Entire books have been written on the subject, and it's impossible to cover it all in just a few paragraphs. If you are interested in more technical detail, two of the best books are *The Solar House: Passive Heating and Cooling* by Daniel D. Chiras (Chelsea Green, 2002) and *The Passive Solar House: The Complete Guide to Heating and Cooling Your Home* by James Kachadorian (Chelsea Green, revised and expanded edition, 2006). In addition, my book, *Natural Home Heating: The Complete Guide to Renewable Energy Options* (Chelsea Green, 2003), has several chapters on passive and active solar home-heating strategies (and a lot of other strategies as well). For additional information on passive solar homes, also see the Organizations and Online Resources section at the end of this book. Now, let's take a look at your next major option—active solar strategies.

Active Solar

The solar domestic hot-water system in my Weybridge home mentioned earlier is one example of the many different types of active solar systems that are available. From my own personal experience I can assure you that a properly designed and installed SDHW system works very well—even in cloudy and cold Vermont. These types of systems are popular in nations around the world, judging by the fact that nearly forty million households worldwide heat their water with solar collectors, most of them installed in the last five years, according to a recent report from the Renewable Energy Policy Network for the 21st Century (REN21) published by the Worldwatch Institute. In the United States, there are about one million residential solar hot-water systems currently installed, according to the U.S. Department of Energy, indicating that there is still plenty of room for additional growth in this sector.

Okay, let's say that like me you live in a conventional house, but want to take advantage of some free solar energy. However, let's also assume that for structural or other reasons a major passive solar renovation just won't work. What are your options? Happily, you don't have to tear the place down and start from scratch. This is where active rather than passive solar heating systems really shine.

In an active system the sun is still the heat source, but a group of specially designed collectors harvest the sun's energy, which is then moved through

A ground-mounted solar-collector array. Courtesy of Solar Works, Inc.

pipes or ducts by pumps or fans to your living space. Most active solar-heating systems have some kind of storage capacity, or heat sink, to provide heat at night and at other times when the sun is not shining. Active systems are divided into two main categories: *domestic hot water* (DHW) and *space heating*. Both of these categories are subdivided even further, with many variations based primarily on the climate zone where they are installed (some simple solar hot-water systems are designed to function without pumps or other mechanical devices, and are referred to as "passive," blurring the distinction between categories somewhat). Both active solar domestic hot-water and space-heating systems contain the same basic components, although a space-heating system has more collectors, a larger storage capacity, and is connected to a home's heat-distribution system. Active domestic hot-water systems are often part of a passive solar house design, while active space-heating systems usually are not, since a passive solar home normally provides most of its own space heat.

Active solar home-heating systems can be expensive to install and will probably not provide for all your heating needs, especially in cold, cloudy northern climates, where a backup heater is necessary. But this brings up an important point: the typical active system is not designed to meet 100 percent of your home heating or domestic hot-water needs to begin with. That's because it would not be cost-effective to try to design a system that would get you through a worst-case scenario of several weeks of cold, cloudy weather in the winter. Rather, active solar is designed to work in combination with other heating systems, which offers quite a lot of flexibility. As long as your expectations for an active solar-heating system are realistic, you will probably be quite pleased with the results. Best of all, active solar-heating systems can almost always be added to existing non-solar-designed homes as a retrofit strategy, making them potentially attractive to virtually any home owner or business with a suitable site.

There are a number of basic design requirements for all active solar-heating systems that are similar to those for a passive solar house. But unlike a passive solar house, there is more flexibility in how these requirements can be met. Ideally, the site where your home is located should have an unobstructed southern exposure to the sun. But even if the south side of your house isn't oriented within 30 degrees east or west of true south, the solar collectors in an active system can be placed on your roof, walls, garage, or even on the ground mounted on a framework at an optimum angle to take maximum advantage of the sun's rays. This flexibility in collector location can make up for a multitude of architectural and site problems that would otherwise eliminate your home from consideration for a solar heating system.

Climate plays a major role in the design and cost of active solar installations. Active solar space-heating systems, in particular, make the most sense where utility rates are high (almost everywhere these days) and in climates that have long heating seasons with a high proportion of sunny days. These systems make less financial sense in areas with short heating seasons or typically cloudy weather patterns. This doesn't mean that you can't install an active system in the Pacific Northwest, northern New England, or Florida, but it does mean that cost-effectiveness may not be your primary motivation. The fact that an active solar home-heating system is more sustainable and reduces the amount of pollution and greenhouse gases emitted by your home or business may be sufficient justification to install one.

Solar space-heating systems are further subdivided into *liquid* and *air* systems. This distinction refers to the heat-transfer medium used between the solar collectors, the storage tank or bin, and (usually) your living space. Liquid systems are more popular than air systems because they generally cost less to operate, take up less space, and are easier to install as a retrofit. Hot-air solar systems are used mainly for space heating, although they are not as popular as they once were, due in part to problems with mold and mildew growth in some storage bins. On the positive side, air systems don't leak or burst. Hot-air systems are most frequently but not exclusively found in drier climates like the American Southwest. Since liquid active systems are so popular, though, let's take a closer look at them.

Liquid Systems

You have many options for different kinds of active liquid solar systems that should match your particular climate, site, house design, needs, and budget. Liquid systems are especially well suited for retrofit projects, but can also be included in new construction. Liquid systems are subdivided into two main categories, *direct* and *indirect*. Direct systems are more appropriate in milder climates, while indirect systems are more popular in colder climates where freezing is a problem. Direct systems use water as the heat-transfer fluid in their collectors and pipes, while indirect systems use antifreeze solutions, or a *phase-change liquid* (such as methyl alcohol) in the collectors and outdoor piping. Although there can be variations, the components in most liquid systems include collectors, a storage tank, pumps, one or more heat exchangers, and controls. The figure below shows these main components (except the controls) in a very simple indirect system. Solar domestic hot-water (SDHW) systems as well as space-heating systems can be either direct or indirect.

A SDHW system heats only your domestic hot water for bathing, washing dishes, laundry, and so on. In a SDHW system, the sun heats the liquid in the collector(s), and the liquid then circulates through pipes into your home. In a direct liquid system, the hot water from the collectors is circulated directly into the hot-water storage tank. In an indirect system like the one in my home, the hot antifreeze liquid from the collector is circulated through a heat exchanger in (or attached to) the hot-water storage tank, where it transfers its heat to the water in the tank. Whether it's a direct or indirect system, the liquid stored in the tank is almost always water.

Liquid solar space-heating systems, on the other hand, store their solar heat in water tanks or in the masonry mass of a *radiant floor system*. As a rule of thumb, most storage tanks need a capacity of about one-and-a-half gallons for each square foot of collector area. The storage tank can be made

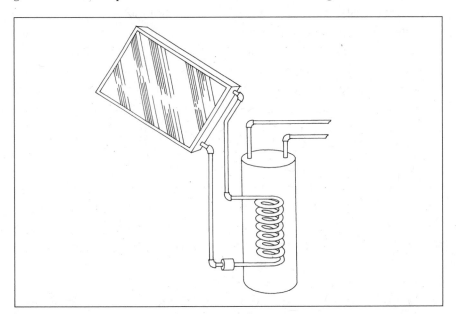

A liquid-active solar-heating system. Courtesy of Michael Middleton.

out of steel, concrete, wood, or fiberglass-reinforced plastic, and should be insulated to reduce heat loss. A distribution system for a liquid solar space-heating installation has two main elements. The first is the piping and pump that circulates the liquid from the collectors to the main heat exchanger and storage tank(s). The second segment includes piping and one or more pumps that distribute the hydronic (hot water) heat from the storage tank to standard heat emitters in your living space, such as radiant floors, baseboards, or radiators.

If you already have hydronic radiant-slab heating in your home, adding a liquid solar-heating system could be an excellent strategy. Radiant-slab heating is the most compatible with liquid solar-heating systems because in addition to storing heat, the slab also performs well at the lower water temperatures (90° to 120°F/32° to 49°C) typical of a solar heating system. What's more, radiant-slab heating is generally more efficient than baseboard heating and tends to "feel" warmer because your body is absorbing the infrared radiation emitted by the slab. Baseboard hot-water heating elements and radiators are not as good a match with solar heat because they typically require water temperatures between 160° and 180°F (71° and 82°C). Whatever type of heat emitters you use, the existing boiler connected to them can serve as your backup heater for times when the sun isn't shining.

While either a solar space-heating or SDHW system can result in substantial efficiency and fuel savings, combining the two systems can result in even greater savings. Either way, the addition of a solar hot-water system to your home offers the possibility of significant reductions in the use of fossil fuels and an increased reliance on a renewable energy source. Most of these solar systems now carry a rating that is similar to the EnergyGuide label required by the Federal Trade Commission for gas and electric hot-water heaters, making meaningful comparisons possible. In the United States, many states and cities now offer a wide range of rebates and other incentives for both businesses and individuals to install a variety of solar hot-water systems. Some municipal electric companies also offer incentives, and in certain cases they will even offer to install entire systems that can be rented or leased.

Santa Clara, California

For example, since 1975 the city of Santa Clara, California, has taken a leading role in the development and promotion of the use of solar energy.

In that year, the city established the nation's first municipal solar utility. Under this program the city supplies, installs, and maintains solar water-heating systems for residents and businesses within Santa Clara. The utility's solar equipment is offered for the heating of swimming pools, process water (commercial or industrial), and domestic hot water. The various elements of hardware (solar collectors, controls, and storage tanks) are owned and maintained by the city under a rental agreement. The renter pays an initial installation fee and a monthly utility fee. These amounts vary depending on the size of the installation.[3] The program has been extremely popular, according to Alan Kurotori, the utility's solar engineer. But the program has also evolved since its inception over thirty years ago. "When the energy crisis died down in the early 1980s, we migrated towards solar pool-heating systems," Kurotori says. "There was a growing demand for them, and we installed over 350 solar pool systems." Nevertheless, the utility continues to maintain its original solar domestic hot-water heating customer base, even after three decades—a tribute to the durability of the systems and the loyal support of their customers.[4] Many other California utilities and municipalities offer a wide range of solar hot-water programs as well.

But California is not the only state to encourage solar hot water. In Hawaii, between 1996 and 2004, more than 25,000 solar hot-water systems were installed by the customers of the Hawaiian Electric Company. These systems have reduced the utility's electrical demand by a total of 12.7 megawatts—enough electricity to power approximately 18,000 typical U.S. homes.[5] It is estimated that there are now over 70,000 homes fitted with solar hot-water systems on the islands.[6] Many other municipalities across the country have similar active solar initiatives, and a number of other states have implemented even larger-scale programs that encourage the use of solar energy—especially for the generation of electricity.

Photovoltaic Systems

Photovoltaic (PV) systems are another increasingly popular strategy for harvesting solar energy. A PV system uses what are called *photovoltaic cells* to convert sunlight directly into electricity. This electricity can then be fed into a circuit and used to power electrical devices or recharge batteries. Individual cells can produce a current of about 0.5 volts of direct current

(DC), the same kind of electricity produced by a battery. A typical PV panel is composed of many solar cells, which contain no liquids, corrosive chemicals, or moving parts. PV panels are rated in watts based on the maximum power they can produce under ideal sun and temperature conditions. A variety of "second-generation" thin-film PV collector materials, in addition to the traditional rigid panel, are now available, and a "third generation" of devices will reach the market soon. Many of these newer collectors can be combined with plastic or other flexible substrates and are less expensive to manufacture. Multiple panels combined together are called an *array*. Arrays can be mounted on the south-facing roof or walls of your home, on various types of pole-mounted racks, or on the ground. As long as sunlight shines on it, the array silently produces electricity, requires very little maintenance, and does not pollute. This makes photovoltaic energy the cleanest and safest method of power generation presently available.

Despite these advantages, PV panels are not very efficient, converting only 12 to 15 percent of solar energy into electricity.[7] Consequently, most PV systems require a lot of panels to produce meaningful amounts of electricity. And since the electricity generated is direct current, a device called an *inverter* is needed to convert the DC power to alternating current (AC)

Photovoltaic panels on ground-mounted tracking racks which follow the sun. Vermont Solar Engineering.

used by most standard household appliances. Inverters capable of producing high-quality waveform output needed by TVs, stereos, microwave ovens, and computers are not cheap. What's more, a number of other electrical control devices (switches, meters, and so on) are also generally needed in most installations, making the entire system fairly expensive. Due to the relatively high cost of most solar electric installations, a solar-powered home must be highly energy efficient in order keep the system affordable. A variety of large-scale incentive programs designed to reduce up-front installation costs and encourage PV use are currently offered in Germany, Japan, the United States, and many other countries.

For a comprehensive listing of information on state, local, utility, and selected federal incentives in the United States that promote photovoltaic and other renewable energy systems, visit the Database of State Incentives for Renewable Energy at www.dsireusa.org. Now let's look at the two main categories of PV systems, *off-grid* and *grid-intertied*.

OFF-GRID SYSTEMS

For many years, most PV systems were used by people who lived "off the grid." These systems were generally considered to be the only viable alternative to prohibitively expensive electric-line extensions in remote locations (and the only option in extremely remote locations). Off-grid systems are designed to be self-contained, and usually include a solar PV panel array, an array DC disconnect, a charge controller, a battery bank (group of batteries), a system meter, a main DC disconnect, an inverter, and a backup generator. The backup generator is mainly needed to recharge batteries in the event of several weeks of cloudy weather, especially during the short days of winter. People with off-grid systems need to adjust their electricity consumption to match their PV system's limits. This does not mean that you have to do without power, but it generally does mean that you have to be more mindful of how much you are consuming—and when you use it. Although these systems are commonly found in remote locations, they will work anywhere; some people just like the sense of security offered by having their own electrical system regardless of where they live. Not having to pay a monthly utility bill is an added bonus. Off-grid PV systems are appropriate for individual homes in small, isolated communities, or for some remote commercial installations.

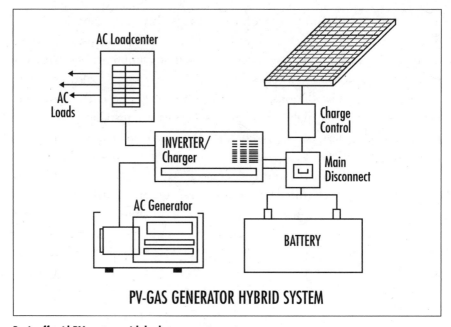

Basic off-grid PV system with backup generator. Rocky Grove Sun Company.

Grid-Intertied Systems

More recently, as various countries and states have passed *net-metering* laws, grid-intertied PV (also known as utility-interactive) systems have become extremely popular, and they now represent the fastest-growing sector in the solar electric industry—vaulting 60 percent a year between 2000 and 2004. More than 400,000 rooftops in Japan, Germany, and the United States now have these systems installed on them.[8] In the United States, about forty states now have some form of net metering laws. Net metering (or net billing) allows PV systems to send excess power not immediately needed in the home directly back into the electrical grid while crediting the homeowner for the excess power. That credit can be applied in months when the PV system produces less electricity than the home consumes. Most net-metering rules allow the homeowner to receive a credit equal to the amount normally paid to the utility for the electricity. Net-metering rules vary from state to state and country to country. In Germany, the homeowner credit is eight times the normal retail charge, largely explaining the wild popularity of PV systems in that country. Ironically, strong demand for PV panels from Germany, Japan, and several states like California and New

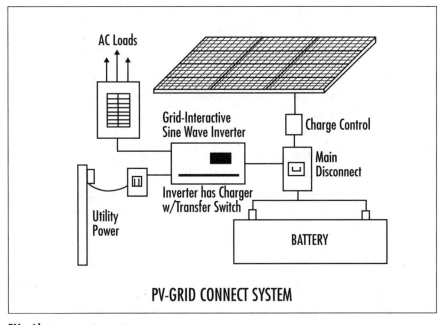

AC Loads

Grid-Interactive
Sine Wave Inverter

Charge Control

Main
Disconnect

Inverter has Charger
w/Transfer Switch

Utility
Power

BATTERY

PV-GRID CONNECT SYSTEM

PV-grid connect system. Rocky Grove Sun Company.

Jersey has led to shortages of PV panels in the past few years, temporarily reversing long-term declines in PV prices.

Most grid-intertied systems include a PV array, an array DC disconnect, and an inverter. Grid-intertied systems can be designed with or without battery backup. Systems without batteries are simpler and less expensive, but don't work when there is a grid power failure. That's why many homeowners choose a system with battery backup, in order to keep critical electrical devices such as a furnace, refrigerator, or well pump running. PV systems with batteries are more complicated, slightly less efficient, and more expensive, but will continue to function for a day or so (depending on the size of the battery bank) during a power outage. A grid-intertied system with battery backup generally includes a PV array, an array DC disconnect, a charge controller, a battery bank, a system meter, a main DC disconnect, and an inverter. Grid-intertied PV systems are a good match for most homes, small communities or neighborhoods, and many municipal and commercial settings. Add the battery backup feature in locations where grid power outages are frequent (which may become the norm rather than the exception in an energy-constrained future).

COOPERATIVE OR COMMUNITY SYSTEMS

When most people think about solar strategies they generally consider using them in their own homes. But cooperative or community systems offer a lot of potential for substantially reducing energy consumption while providing convenient, reliable domestic hot water, space heating, or electricity for larger groups of people in a neighborhood or community setting. This approach is fairly routine in some European countries, but less common in the United States, although this is beginning to change with an increasing number of community-supported energy initiatives in some locations. This approach could—and probably should—be a primary strategy in community responses to peak oil everywhere.

Westwood

One of the best examples of this strategy in this country can be found at Westwood in Asheville, North Carolina. Completed in 1998, Westwood is a residential urban eco-neighborhood and intentional community located on four and a half in-town acres. Its buildings are among the top 1 percent for energy efficiency in the United States, according to a U.S. Department of Energy database. But Westwood is also a beautiful and pleasing place to live, with twenty-four clustered townhouses; a large community building that offers many shared facilities and resources for community members; and gardens, playgrounds, pedestrian paths, and woods, all integrated with the housing. Parking is located at the edge of the community, placing primary emphasis on living spaces for people rather than vehicles.

At the heart of Westwood is a community building with its central thermal-solar collectors mounted on the south-facing roof, and a large solar hot-water storage tank in its lower level. The tank feeds a sophisticated central (district) system that distributes warm and hot water to all buildings for radiant-floor space heating and domestic hot water through an underground network of pipes. Bill Fleming, the engineer and Westwood codeveloper, designed the mechanical and electrical systems for the community. "I had been aware of resource depletion for decades, both in the United States and abroad," he says about the thought process behind the early planning that went into the community's design. "So it was natural

Solar domestic hot water collectors on the roof of the community building at Westwood CoHousing Community, Asheville, NC. Shelter Technology, Inc.

for me to promote the concept of a very, very energy-efficient campus. A number of strategies were involved, and solar was one of them."[9]

But Fleming stresses that he and the others involved in the process were working with both sides of the energy equation—supply and demand—and that both were considered to be equally important. By significantly reducing the design heat load of the 30,000 square feet of buildings on the site, they were able substantially to increase the potential solar contribution. "That required working with the architect, the general contractor, and the trades, in making the buildings extremely energy efficient," Fleming recalls. The result was a group of structures that approached Canadian air-tightness standards—in North Carolina. The buildings also subsequently earned a score of 93 in the Energy Star rating system, approximately double the threshold for Energy Star rating, and about 60 percent better than the local building code, according to Fleming.

On the supply side of the energy equation was the district solar hot-water system. Fleming had built solar systems on houses before, but he decided that he did not want to install individual systems on the clustered townhouses at Westwood, which were arranged for community interaction and aesthetics, rather than for solar orientation. "Solar orientation is not necessarily compatible with those other criteria," he says. "So, we oriented the community building for optimum solar gain along with an appropriate roof pitch. Then, we designed a support system for the large and heavy solar array and worked with the general contractor to get it built correctly." On the lower level of the community building, adequate space was designed for a 1,000-gallon insulated water storage tank and its associated pumps and controls. The tank holds about 1 million Btus of accessible heat and warms to about 150°F (65°C) on a sunny day. "It's always there, tempering every bit of hot water and contributing to the space heating," Fleming says. "But most important of all, we were working with total system design right from the beginning." Part of that original design included central, natural gas-fired water heaters to provide backup heat. Plans are underway to add a wood-fired backup boiler to the system in light of continuing increases in natural gas prices and concerns about long-term supply.[10]

Overall, the solar hot-water system has performed extremely well, according to Elana Kann, Westwood's codeveloper. A special energy-saving feature has been particularly effective. "At night, when nobody is using the domestic hot water and when the radiant-floor heat is not needed, the system goes to sleep," she explains. "It stops circulating, and the heat is retained in the central storage tank in the common building. But it responds to demand. There is a small demand button next to every water tap, and all you have to do is push the button and the system begins to circulate again. Then, if nobody uses it, the system shuts itself off again. We don't know how to quantify the savings resulting from that, but I'm sure it is significant."[11]

Of course, the system does need occasional maintenance, and a team of Westwood residents generally keeps an eye on it. However, these people are not mechanical professionals and they also tend to be busy with other activities, a situation that has occasionally led to some lapses in maintenance. "The next time we do a community like this it's going to be three times the size so we can afford at least a part-time manager on site to keep an eye on things," Kann says. "In order to do that, I think we would need seventy-five to one hundred dwellings."

Obtaining permit approval for the community—and especially the central district-heating system—turned out to be surprisingly easy, according to Kann. "The system was new to the permitting officials and the inspectors, but Bill [Fleming] was wise enough to put together detailed plans and met with every department that it affected and satisfied their concerns before we submitted the overall plan. By the time it came to the actual permitting, it just sailed through. The Asheville planning officials were moving in the direction of sustainability anyway, and they were really hungry for developers' initiatives that made sense. They were absolutely delighted with the proposal and it passed unanimously before the planning board."

The project did run into other problems, however. "There were two main difficulties that we encountered which almost stopped the project," Kann admits. "One was internal in the group. This is why we no longer build cohousing communities; we now say that we use 'co-housing principles.' That's because along with the term 'co-housing' there is an assumption that the group is the developer, and there was massive confusion about this. The buyers had a lot of input, but when we started construction, some people thought they were in charge, and an amazing power struggle developed. We finally just had to pull away from the group in order to get the project done. We thought we had been clear about it all." Unfortunately, for several years afterward there was a lot of contention in the community resulting from the misunderstanding.

The other issue was financing. "When we went to get construction loans, we wasted about six months negotiating with three or four different banks," Kann continues. At the time, we only had two houses left that had not been pre-sold. In every case the local bankers would say 'this looks fine.' Then, they would take our application to the main bank and would come back and say that they needed 'just one more thing.' And each time we kept doing that, they kept coming up with yet another request. Finally, we realized that they just didn't want to do it. So, we gave up on the banks and formed our own bank, and put together a loan pool."

The strategy worked, and the project moved forward. "A lot of the buyers put up money," Kann says. "Those who could afford it bought their houses ahead of time, basically as loans, so at closing, instead of them giving us money, we gave them interest on their loans. There were two buyers who were able put in a lot more than that, and there were a number of interested bystanders in the community who wanted to see us succeed who just loaned us the money. By doing very detailed financial modeling we were able to borrow a minimum amount of money and then plow it back in at the first set of closings. As the first people moved in, we put the money into more construction in a second phase so we did not need the full amount of money up front. We had to take it out of the banks' hands and follow a different route, but it worked out very well for us."

Kann has some advice to others who might be considering a similar project. "Get a developer who has experience involved right from the beginning," she says. "With an experienced developer on board you'll have large parts of the knowledge necessary to put something like this in place. It's almost impossible for an inexperienced group to pull this sort of thing off." Kann also suggests working with people who are knowledgeable about renewable energy systems. "I hesitate to say 'experts,' because there are an awful lot of people who claim to be 'experts' who give very bad advice in this area right now," she notes. "However, without hesitation, I recommend a central heating system. It just doesn't make any sense to build a house on a lot with its own system any more. Don't even consider doing a project with a typical box on a lot; from an energy standpoint that's a ridiculous thing to do, especially knowing what we know now. Cluster those houses together with shared walls and radiant floor heating, use lots of insulation, caulk, and spray foam to make sure they're tight. Also, do it on a scale that can support the necessary maintenance staff. Smaller clusters of houses are great for social reasons, but the total number of households probably should be between seventy-five and one hundred in order to support the central services infrastructure and maintenance personnel."[12]

Sebastopol, California

In California, about fifty miles north of San Francisco, the city of Sebastopol (population 7,500) has followed a slightly different cooperative solar strategy that has nevertheless attracted quite a lot of attention both in this country and abroad. In March 2003, the city entered into an agreement with Cooperative Community Energy (CCEnergy) to install 1 megawatt of solar power within the city limits over the following two years through a program called Solar Sebastopol. The program was designed to make the purchase and installation of solar PV equipment easy and affordable for residential, commercial, and municipal customers. CCEnergy pro-

moted, implemented, and administered the program by conducting outreach and education activities, arranging financing options, cultivating a
network of qualified local solar designers and contractors, and facilitating
bulk purchasing of solar equipment at discount prices.

"Solar Sebastopol is an extension of the cooperative model that CCEnergy
has developed to help people to help themselves," CCEnergy president
Daniel Pellegrini said at the start of the program. "It's nice to see a community as a whole embrace solar energy. We want to facilitate making this
dream come true."[13] CCEnergy is a San Rafael–based solar energy cooperative that buys solar equipment at volume discounts, connects solar purchasers
to qualified contractors, and provides a wide range of project management
services for solar installations. Sebastopol's agreement with CCEnergy was
the culmination of a feasibility study conducted in 2002 by Sonoma State
University Energy Management Design Program students during a time of
steep utility rate increases. The students mapped and rated properties within
city limits and assessed the city's ability to secure financing to determine
Sebastopol's overall solar energy potential. The Sebastopol City Council
endorsed the study's conclusions and began seeking a consulting partnership
to implement a citywide solar energy program. CCEnergy was selected.

In 2003, the city council approved the installation of two 10-kilowatt
photovoltaic (PV) systems on the roof of the public works building and the
fire station. Since then, the systems have performed well beyond expectations and have saved the city over $10,000 while providing a clean source
of electric power. The city had originally estimated a fourteen-year payback
on the net cost of $92,000 for both systems, but it now expects to realize
the savings in less than eleven years. In June 2004, the city provided
funding to a local public swimming-pool association for the installation of
a 17.6-kilowatt PV system. Anticipated savings from reduced electric bills
are expected to pay off the financing expense in about ten years. In August
2005, a 28.3-kilowatt solar system was completed on the roof of the city's
high school. The system is expected to produce 55,544 kilowatt-hours of
electricity annually, providing about 15 percent of the school's needs. In
addition, a large number of individual residential and commercial systems
have also been installed in the city and surrounding area.[14]

Sebastopol mayor Larry Robinson has been instrumental in moving the
city toward greater energy independence—and in raising public awareness

Solar photovoltaic panels on the roof of the Analy High School in Sebastopol, CA. Lori Houston, Cooperative Community Energy.

of peak oil issues. Robinson says that the problem with large-scale, centralized, utility-based electricity generation is that "somebody else is in control of our energy. Whereas a PV system on your rooftop or a small wind generator or a community-owned generator puts the power literally in peoples' own hands, and that's both a more democratic way, and I think ultimately safer and more sustainable."[15]

Despite its impressive gains, the program has not met its original installation goal. "We have not moved as quickly toward our goal of 1 megawatt of solar power constructed in the city limits by the end of 2005," Sebastopol city manager Dave Brennan noted, "but we will be over 300 kilowatts within the city and probably an equal or greater amount constructed around the city by the end of the year."[16] This shortfall is due to a variety of unforeseen hurdles. Most important was the fact that solar PV manufacturers were simply unprepared to meet the sharp increase in demand in North America caused by the popularity of various state and local incentive programs; during 2004, the demand for PV panels far outstripped supply. This situation squeezed profits for smaller installers and companies, forcing many out of business, making a bad situation even worse. Tight supplies of PV panels continued throughout 2005, despite the fact that solar panel manufacturers invested heavily in new production facilities, and the backlog of pending solar projects in Sebastopol grew larger.

Another unanticipated problem concerned the city's intention to offer low-cost financing for the nonrebate portion of the system costs. In recent years, Sebastopol has experienced severe budget shortfalls, and the city decided it could not provide public funds for private gain. Nevertheless, the city continues enthusiastically to support the program in many ways and has also lowered building permit fees for residential solar systems, while adding an additional permit fee for projects that do not include solar. Ironically, the very success of the state's solar rebate programs created budget problems that delayed repayments and even threatened to eliminate rebate funding. All of these unforeseen developments substantially slowed the pace of the Solar Sebastopol installations. Despite the many challenges, however, virtually everyone involved considers the program to be a success. Solar Sebastopol has unquestionably promoted reliable, self-generated solar energy, and has provided a model for other communities across the nation to follow.[17] (For additional information about the Solar Sebastopol program visit www.solarsebastopol.com.)

Cooperative Community Energy

Cooperative Community Energy, which played a key role in the first two years of the Solar Sebastopol initiative, provides an outstanding model that could be replicated elsewhere, and consequently deserves a closer look. CCEnergy grew out of an Earth Day Solar Energy Forum held in Fairfax, California, on April 22, 2001. The forum happened to coincide with the worst energy crisis in California since the oil shortages of the 1970s, and interest in renewable energy in general—and solar energy in particular—was intense. Dan Pellegrini, the president and cofounder of CCEnergy recalls the fair. "There were hundreds of people there," he says. "But we realized that if we wanted anyone to help us with renewable energy we were going to have to do it ourselves."[18] The co-op evolved from a series of meetings and investigations that took place in the months that followed the forum, which identified the desire of area residents for lower prices on renewable energy equipment as well as quality installation services.[19]

The group's initial strategy was to organize a bulk purchase of photovoltaic equipment. The idea was to combine the demand for system components to take advantage of volume-discount pricing and to provide leverage in price negotiations. At first, the group tried to work with a local distributor to place a large equipment order, but the results were disappointing. The group then widened its search for equipment and suppliers, but eventually decided to deal directly with equipment manufacturers. In this strategy the group was acting as wholesaler and distributor, and needed to create a business entity to process the orders. Cooperative Community Energy was born. But why a co-op?

"We looked at various business models, and kept coming back to the cooperative model as being the best way to do it," Dan Pellegrini says. "We felt that it was important to have an organization that was democratically run, that was completely transparent and firmly on the side of consumers. We had an opportunity to build an organization that we, as consumers, would like to see. As a cooperative, it's a nonprofit, member-benefit organization, and from a social perspective it just seemed to be the right way to organize ourselves as a community-based organization."[20]

There was a lot of early discussion about also providing installation services, but CCEnergy decided to assist in the selection of qualified installation professionals in the local community. Since its inception, the co-op has been involved in over 275 solar projects, primarily (but not exclusively) in the San Francisco Bay area. Over time, the co-op has come to appreciate that the long-term success of renewable energy depends on strict compliance with rigorous engineering standards, and it has been building processes to ensure a high level of quality for the installations done for its members. "We made an early decision to only provide the best equipment and service, and we tried to set a benchmark in an industry that had no great interest in doing that up until then," Pellegrini says. "Three or four years later, the bulk of the solar providers now are providing the sort of things that we were providing to our members pretty much from the beginning. We're not the only reason that's happened, but it's extremely satisfying to see that taking place."[21]

Currently the co-op's primary focus is purchasing solar photovoltaic and thermal systems and managing residential, commercial, and municipal solar installation projects. "Initially, we tried to stay focused on PV because we wanted to get really good at one thing, rather than trying to be a jack-of-all-trades," Pellegrini says. Now that it has established that expertise, the co-op is broadening its scope of activities to include energy efficiency and consulting services. CCEnergy is also beginning to help members purchase the means to harness other forms of renewable energy, such as wind, and has already lined up a number of purchasing agreements with wind-turbine manufacturers.

"This started as a group of homeowners who were concerned about solar energy for their own use, but it's grown far beyond that now," Pellegrini says. "We're not just doing residential properties anymore. Now we're involved with much larger commercial properties, agricultural projects, schools, and nonprofits, and we're doing a lot in affordable housing, churches, and so on. We don't want solar energy to be just for rich white folks in Marin County; we want to democratize it and make it available to a much wider group of people. We're trying to provide some kind of finance mechanism where people can invest in solar energy without having to face the huge upfront capital costs. I'm working with a local group that has been trying to develop various ways of financing renewable energy projects, with our focus on financing smaller solar energy projects and not just the big megawatt projects."[22]

Muir Commons
One of the largest residential systems that CCEnergy has been involved with was installed in June 2002 at the twenty-six-residence Muir Commons Co-Housing Community in Davis, California. Muir Commons is a good example of the co-op's expanded focus beyond individual residential projects. One of the more unusual aspects of the project was that a five-day workshop was part of the system installation. CCEnergy, Muir Commons Co-Housing, and Sine Electric of Santa Rosa collaborated on the hands-on, photovoltaic solar-panel design and installation workshop. Designed for both homeowners and contractors, the workshop included content and materials applicable to both residential and small commercial installations anywhere within the United States. The fifteen workshop attendees came from Canada, Florida, Colorado, Utah, Southern California, and the Caribbean.[23]

In addition to providing information and resources, the workshop included specific skills development and practical application: workshop attendees participated in the actual installation of a ninety-six-panel, direct utility-intertied solar PV electric system at Muir Commons. Beyond the distinction of being the first co-housing facility in the United States developed from the ground up, the newly installed 10-kilowatt PV system has made Muir Commons the largest residential producer of solar energy in the city of Davis.

Steve Lyons, owner of Sine Electric, was the project manager. "The community has a common building that is used for community activities," he explains. "It contains a commercial-grade kitchen, a laundry facility, a hot tub, an office, and a daycare center, as well as a fairly large common activity room. The common building is composed of three conjoined structures, each one of which has its own air-conditioning system. The calculations for offsetting utility use were done by CCEnergy, and it was determined that a 10-kilowatt PV system would offset about 90 percent of the historic use of that structure's electrical services."[24]

As it turned out, the system performed even better, owing to an unintended consequence of the design and installation, according to Lyons. Because of the large number of solar modules that were installed on one of the roofs, they reduced the solar gain in that building so much during the summer that only one-third of the air-conditioning was needed as before, further reducing the overall electrical load. "They are now net producers of electricity in that building, and all of the energy that was used for that structure has now been offset," Lyons adds. Although the combination of a workshop with a fairly large-scale installation offered some management and logistical challenges, overall the project went well. "It was especially wonderful to see the enthusiasm of the participants," Lyons says of the project. Dan Pellegrini agrees. "It was a wonderful class," he adds. "The participants learned a lot, and we ended up with a really nice installation."

Solar Energy International

Some of the curriculum for the workshop at Muir Commons was adapted from a comprehensive body of instructional materials developed by Solar Energy International located in Carbondale, Colorado. Founded in 1991, SEI is a nonprofit organization whose mission is to provide education and technical assistance to empower people to use renewable energy technologies. Since its inception, SEI has grown steadily. The organization now has approximately a dozen employees, and regularly holds training sessions in about a dozen different sites across the country and a number of other locations around the world. Workshop subjects include solar electricity, wind power, micro-hydro, solar home design and natural house building, solar thermal, renewable fuels, and a special series of courses for women.

Along the way, the group's focus has evolved over time, according to Johnny Weiss, SEI's cofounder and executive director. "Historically, we served a very broad population of people that wanted to get involved with renewables and wanted to bring these technologies into their own lives and families," he says. "But, in the last decade, that's changed quite a bit. Now, we're serving people who want to get involved in careers in renewable energy and natural house building. There has been a bit of a shift, but it's still very grassroots and hands-on focused. There's also no question that photovoltaics have become an increasingly large part of our work. We published a textbook on photovoltaics in 2004, and PV now represents about 60 percent of the training that we do."[25]

One recent PV project that SEI was involved in that had a strong grassroots community focus took place in Nevada. In April 2005, SEI worked with the Battle Mountain Band of Te-Moak Western Shoshone to provide free training and installation of a solar photovoltaic system in the heart of Western Shoshone territory near Elko, Nevada. Additional partners in the project were Honor the Earth, the Western Shoshone Defense Project, and American Spirit Productions. The solar panels and energy-efficient systems were installed to power the ranch home of Mary and Carrie Dann, two Western Shoshone grandmothers who have been central to the Western Shoshone quest for environmental justice, treaty rights, and sovereignty for more than forty years. Participants in the training received certification in solar energy installation, which is the first step toward the promotion of locally run energy systems and economic development for Native communities. The installation also acted as a model of alternative energy development in a desert region whose tremendous solar potential has been largely untapped. Most importantly for the Shoshone, the project serves as a beacon of safe, clean energy in an area slated for housing the high-level nuclear waste repository at Yucca Mountain, Nevada. The Western Shoshone Nation, the state of Nevada, its residents, and its congressional delegation all overwhelmingly oppose the dump and have fought vigorously against it for decades.[26]

"They are getting very excited about renewable energy opportunities and doing something positive and taking control of their own energy future," Johnny Weiss says about the Western Shoshone project. "We assisted them in implementing a renewable energy project and training opportunity for people in the tribe to learn about solar electricity as well as the installation of a working system. As part of the project, we did classroom and lab exercises, but then we went out with the students and installed a two-kilowatt system that was the only one on tribal lands that is a working demonstration of solar electricity. The students and the community were enthusiastic, and it was both a pleasure and a privilege to work with them."[27]

Speaking of community, SEI has also been involved in a solar project right in its own hometown of Carbondale, where a grid-connected PV system was installed on the town hall in August 2005. Participants in a PV design and installation workshop helped Carbondale have one of the few town halls in the country powered mostly with photovoltaics. The project was a collaboration between the town of Carbondale, the Community Office of Resource Efficiency, Sunsense Inc., SEI, and Xcel Energy. "We have a very supportive local government that wants to rely more on renewables, and with the help of our local utility company we've been able to install a four-kilowatt PV system on the town hall," Johnny Weiss says. "Everybody in town benefits from the lower utility bills on the town hall, and it also serves as a great demonstration and represents the community's commitment—and particularly this town council's commitment—to trying to do something positive about our energy use. They're actually doing it themselves and have invested quite a lot of time, money, and effort into making it happen."[28]

But that's not all. SEI has brought the grassroots model of community involvement into the heart of local communities—their schools—with its Solar In the Schools (SIS) program. The SIS program has its roots in a camp that was organized about four years ago to teach Native American youth about renewable energy. Since then, the program has shifted to spreading the educational message to the wider community. In the spring of 2004, SEI presented its renewable energy curriculum to over 600 students in fifteen different classrooms around Colorado, culminating with an installation of a 1-kilowatt PV system on the Basalt Middle School in Basalt, Colorado. In 2005, the program continued to gain momentum as teachers all around the state (and country) contacted SEI to find out more and asked SEI to come to their classrooms to give renewable energy presentations. One of the most successful collaborations took place in Crested Butte, Colorado, during the week of September 19 between SEI, the Crested Butte Office of Resource Efficiency, and the Crested Butte Community School. The local utility, Gunnison County Electric Association, suggested the idea, and donated $10,000 to get the project rolling, while BP Solar donated the PV panels.

The highlight of the program was the installation of a 1.55-kilowatt PV array in addition to hour-long, hands-on presentations to each grade. Students from the fifth, eighth, and tenth grades each spent an hour putting the rack and solar array together. Some of the kids got to test the panels, others assembled the rack, while the students from the older grades wired the panels together and put the finishing touches on the array. The installation was followed two weeks later by a solar fair, a day-long event held on the school grounds to further educate the community while showcasing the new PV system. "The kids loved it," says Soozie Lindbloom, coordinator of SEI's Solar In the Schools program. "They came to all the

The fifth graders at the Crested Butte Community School and the components of their new 1.55 kW PV system. SEI staff.

displays about different renewable energy technologies, and the older kids just took over and started teaching the younger kids and the community members that came to the fair. We listened to what they were saying and it was what we had taught them—and more."

The ability to learn about renewable energy and then actually implement a working system (complete with a data-monitoring component) has been an empowering experience for the students that they have been sharing with their parents. "I would be walking around town, and the kids would see me and say, 'It's the solar girl!' and then proceed to tell their parents everything they learned from SEI," Lindbloom says. "This was one of the best projects I've ever been involved with. SEI came into the school and really created a huge spark that has already started to ignite that educational fire—that desire to learn more about renewable energy. All of the teachers were really excited to receive their curriculum kits and start implementing more renewable energy education into the classroom. The students were just amazed by the simplicity, the elegance, and the independence of renewable energy technologies. Crested Butte is a small, isolated mountain town, and it was fascinating to watch the information that we brought to the school spread throughout the community—it was simply amazing."[29]

CHAPTER THREE
WIND POWER

In the early 1980s, I bought a house located in the middle of ninety acres of remote forestland in Lincoln, Vermont. The nearest electric line was half a mile away, and I soon discovered that it would be prohibitively expensive to have the line extended to the homesite. I considered my options. Photovoltaics were simply too expensive at the time. But the house was located on a hillside that offered fairly good wind potential, and after a lot of research I decided to install a Sencenbaugh SWE-1000 wind turbine[1] on an eighty-foot, guyed steel tower. Jim Sencenbaugh, the Palo Alto, California, manufacturer of this battery-charging unit, had a good reputation for producing a high-quality, well-designed product, and even today his machines are highly regarded and considered a real find in the used turbine market.

The turbine arrived, partly disassembled, in a group of heavy shipping crates—complete with assembly instructions. After a lot of preparation and waiting for the weather to cooperate installation day finally arrived. I had agreed to assist the experienced installer in the process, but was a bit surprised to discover that my role was to climb the tower to help place each new section of tower, bolt it in place, and connect the guy cables at appropriate intervals. As the tower got taller, I grew increasingly queasy on the gently swaying, unguyed sections—even though I was wearing a safety harness. But I hung in there, and by the end of the second day the wind turbine was installed atop the now completely anchored tower. From the ground, the view of the mountains marching off to the misty horizon in the south was beautiful. The same scene from the top of the tower was simply breathtaking, especially during fall foliage season.

As the years went by, I had many opportunities to take in that view, as there was quite a bit of routine maintenance that needed to be performed on the turbine. I learned to plan every climb carefully, and tried not to forget (or drop) anything, as each climb up and down the tower was a time-consuming and tiring operation. I also learned to stay off the tower during windy or stormy conditions. Overall, the system performed well. The battery bank in the house powered by the turbine provided modest amounts of electricity for the few lights and small appliances I used, and generally met my equally modest expectations. I quickly learned to match my energy consumption with the available supply. Even after I sold my Lincoln home, the wind turbine provided fairly reliable power for many years.

THE WINDS OF CHANGE

Wind turbines have changed dramatically since they were first developed by Poul la Cour, a Danish "Edison," in 1891. Early wind turbines were capable of generating a few kilowatts, enough to power a house or two. Now they are hundreds of times larger. Commercial-scale turbines generally range from about 500 kilowatts to 3 megawatts (3 million watts) in generating capacity—or even more. The most powerful is a 5-megawatt giant, the REpower 5M located in Germany, with a rotor diameter of 413 feet, which is longer than a football field. In 2004, global wind capacity reached nearly 54,000 megawatts, according to author and wind analyst Paul Gipe (that figure increased to over 59,000 megawatts by the end of 2005, according to the Earth Policy Institute). From 1990 to 2005, wind has been the fastest-growing power source worldwide on a percentage basis, with an annual average growth rate of about 29 percent in the past decade. Wind power installations are primarily concentrated in Europe and the United States, but can be found in many other nations as well. As of January 2006, over 9,150 megawatts of wind power capacity was operating in the United States, generating enough electricity for about one and a half million average American households, Gipe says. The cost of wind-generated electricity has declined nearly 80 percent since the development of modern wind turbines in the 1980s—making it an attractive generating option today. And because wind power uses a local resource as "fuel" rather than imported forms of energy, it can have a more favorable economic impact than investments in coal or natural gas, according to a recent study by the National Renewable Energy Laboratory.[2]

A LONG WAY

The U.S. wind industry has come a long way from the early 1970s when the only turbines available in the United States were small, 1930s-era units salvaged from Western farms and ranches. Since then, there have been many steady advances and refinements in wind technology that have resulted in much more reliable machines. Today there is a wind turbine on the market for almost every use, from tiny 20-watt fence chargers with 0.5-meter[3] (20-inch) diameter rotors to 3-megawatt giants with rotor diameters that approach the length of a football field. In between those extremes are many possible choices.

Wind turbine designs fall into two main categories, *horizontal-axis* (propeller-style) and *vertical-axis* (so-called eggbeater-style) machines. Horizontal-axis wind turbines are by far the most common, although there have been some interesting experiments with vertical-axis machines. Turbine subsystems include a *rotor*, or blades, that convert the wind's energy into rotational shaft energy; a *nacelle* (enclosure) containing a drive train, sometimes including a gearbox[4] and always a generator; a tower to support the turbine; and electronic equipment such as controls, electrical cables, and interconnection equipment. Off-grid turbines do not have interconnection equipment, but normally do

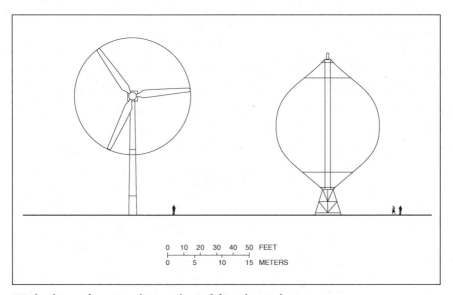

Wind turbine configurations: horizontal axis (left) and vertical axis. Pacific Gas & Electric.

have banks of batteries to provide electricity storage for windless periods. Since the rotor is what actually captures the wind, its size is extremely important, and in general the larger the rotor the better, as long as it is matched to an appropriate-sized generator.

Wind turbines can be roughly divided into three main categories based on their size: small, medium, and large. Small turbines are frequently subdivided further into *micro*, *mini*, and *household* size machines. The micros typically have rotors that are less than 1.25 meters (4 feet) in diameter, mini (or cabin-size) turbines have rotors less than 3 meters (10 feet) in diameter, while household machines can have rotors up to 8.8 meters (29 feet) in diameter. In most cases, small turbines are used by single households, although they may power isolated communities with limited electrical demand in some developing nations. About three-quarters of all small wind turbines end up in stand-alone power systems at remote sites.

Medium-sized turbines are larger. Some evolved out of the commercial turbines of the early 1980s with rotor diameters from 10 to 15 meters (30 to 50 feet), while others now approach megawatt size with rotors between 50 to 60 meters (150 to 200 feet) in diameter.[5] The final turbine category, large, includes machines in the 1- to 3-megawatt range. These are the giants that are typically used in large-scale, commercial wind farms in North America and elsewhere. Medium-sized and large machines offer significant opportunities for groups of families, neighborhoods, or communities to provide significant amounts of their own electricity. Since they are especially well suited for cooperative or municipal installations, medium-sized and large wind turbines will be the primary focus of this chapter. If you want to install a smaller turbine for your own family, that's fine too—as long as you have a suitable site with good wind resources.

LOCATION, LOCATION, LOCATION

Regardless of its size, selecting a good location for a wind turbine is extremely important. In most cases, careful testing of wind conditions with an *anemometer* at a proposed site for at least a year is vital, since the most important factor in the amount of power available to a wind turbine is the speed of the wind.[6] And because the power in the wind is a cubic function of wind speed, even minor changes in speed can have a profound effect on power output. There are a variety of factors that tend to influence wind speeds. Winds tend to be stronger and more frequent along the shores of large lakes and the ocean because of the differential heating between the land and water. Mountain-valley breezes are another example of local winds caused by differential heating. Mountain-valley winds can be reinforced when prevailing winds flow in the same direction; Altamont Pass in California is a good example of this phenomenon. At a windy site, a wind turbine will generally be in operation about two-thirds of the time, or about 6,000 hours per year. However, even with a good location, a wind turbine may not perform at its best if the tower is not sized properly. Experienced wind experts, such as Mick Sagrillo,[7] say that the three most common mistakes people make with wind installations are:

1. too short a tower,
2. too short a tower, and
3. too short a tower.[8]

Today, a tower height of at least 80 feet (25 meters) for household-sized turbines is considered to be the minimum.[9] If there are trees nearby, 100 to 120 feet (30 to 35 meters) is appropriate. Commercial-sized wind turbines often use towers that are much taller. "There are companies in Germany that are putting wind turbines on 160-meter (525-foot) towers, and they have to have an elevator because you can't get any work done if you try to climb it," says Paul Gipe, who is the author of numerous wind power books and articles.

Towers come in two main categories: freestanding and guyed. Freestanding towers require large and deep foundations to prevent the tower from tipping over in high winds and must be internally strong enough to withstand strong gusts. Guyed towers, on the other hand, use a smaller foundation and a number of anchors and cables to achieve the same end. Guyed towers are the most common choice for small wind turbines. Freestanding towers are more expensive, but take up less space than guyed towers. Freestanding towers are further subdivided into *truss* (or lattice) and *tubular*. These days, nearly all medium-size and large wind turbines are installed on tubular towers.[10]

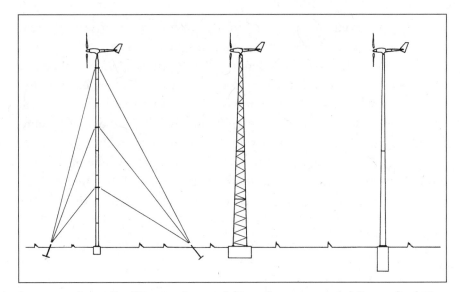

Tower types. For small turbines, guyed towers (left) are the most economical. Freestanding lattice (center) and cantilevered tubular towers (right) are more costly. Bergey Windpower.

Ironically, one of the greatest advances in off-grid wind technology is only indirectly related to wind—photovoltaics. Back in the 1980s, there wasn't much cooperation between the wind and photovoltaic industries. Each touted the benefits of their respective technologies, but generally failed to recognize the potential benefits of combining the technologies in ways that would enhance their respective strengths and minimize their weaknesses. In many locations, the strong winds of winter combined with the long sunny days of summer can be utilized most efficiently by a hybrid wind and solar installation that allows the designers to reduce the size of each part. Small backup generators provide even more flexibility and can reduce the size of the battery bank needed on off-grid systems. These types of hybrid systems are an excellent choice for any off-grid location, and are especially popular in developing nations where extending the (frequently unreliable) electrical grid into remote locations can be problematic or pro-hibitively expensive.

There are many books on wind power, but the best is *Wind Power: Renewable Energy for Home, Farm, and Business* by Paul Gipe (Chelsea Green, 2004). Anyone interested in wind energy, regardless of the size and scale of the system, will find this book invaluable, and I highly recommend it.

ANOTHER MODEL

During the OPEC oil embargo and subsequent oil crises of the 1970s there was a flurry of wind-turbine activity in the United States, and some significant advances were made in the technology. But in the early 1980s, when the Reagan administration dismantled the energy tax credits and incentives that had encouraged the installation of alternative energy systems nationwide, the U.S. wind-power industry collapsed. Denmark, on the other hand, understood the incredible long-term potential for wind power. In 1980, a newly elected Social Democratic government offered a 30 percent subsidy for the construction of new wind-energy projects. This initiative had its roots in the Danish antinuclear movement, but eventually it expanded to become a positive model for alternative energy production. After a twenty-year partnership between government and industry, Denmark emerged as the world leader in the wind industry.

What most Americans don't know, however, is that the vast majority of wind installations in Denmark are composed of small groups or clusters of mid-sized turbines, not the huge wind farms found in North America. And the vast majority of these Danish wind turbines are operated by farmers, homeowners, and small businesses, either independently or (more frequently) as cooperative ventures. This same general cooperative strategy can also be found in wind-turbine installations in other northern European countries, especially Germany, the Netherlands, and Sweden. These countries have clearly shown that medium-sized wind turbines can be used to power farms, homes, and businesses at a scale somewhere between the small individual homeowner installation and the large-scale commercial wind farm of today.

This community-supported energy model offers a lot of possibilities in North America, especially for communities that are trying to provide a greater degree of energy security or self-sufficiency in an era of increasingly uncertain energy supplies. Yet there have been problems in transplanting this concept across the Atlantic Ocean. "The potential for community wind is huge, but there are many obstacles in North America," Paul Gipe says. "Utilities everywhere resist this, and virtually everything is stacked against you. The first problem is that you can't connect to the grid. Even if you can connect, you can't get paid. And even if you can get paid, you can't get paid enough. Other than that, there are the typical problems that you have with any development, such as financing, siting issues, permits, zoning approvals, and so on."[11]

Despite these many challenges, community wind has a lot going for it, according to Gipe. Unlike traditional, large-scale power plants, wind energy is modular, meaning that each wind turbine is a self-contained power plant that can be sized and located almost anywhere. Wind turbines can be grouped together in large clusters or dispersed widely across the landscape. And wind turbines can be located where the power is needed, reducing the need for heavy upgrades of high-voltage, long-distance transmission lines. What's more, wind-energy projects can become a key component in local economies by offering farmers and area residents the opportunity to build, own, operate, and profit from their own renewable energy systems.

Another advantage of community wind is that it can be owned individually, cooperatively, or collectively through a variety of legal mechanisms. If one strategy doesn't work, another might. Although the ownership strategies can include limited liability corporations, cooperatives, school districts, municipal utilities, or combinations of these models, what they all have in common is some form of community ownership and group benefit. The main point is to identify the project as belonging to the community, which may avoid (or at least minimize) the usual conflicts between local residents and developers, whose large-scale, commercial proposals are often viewed as primarily benefiting absentee owners.

Municipal utilities, in particular, offer fertile ground for community-owned wind projects in the United States. There are approximately two thousand community-owned electric utilities in this country serving over forty-three million customers in small, medium, and large communities in all the states except Hawaii; however, 70 percent of these utilities serve communities with a population of ten thousand or less. Public utilities' installed generation capacity represents about 10 percent of the national total (while co-ops account for about an additional 4 percent). Because municipal (public power) systems in the United States are locally owned, they tend to be responsive to their customer-owners. In recent years, these customers have been asking their public power providers to be more environmentally responsible in their choice of power-generation sources.

Consequently, public power systems have achieved substantial reductions in greenhouse gas emissions and other pollutants, and lead the industry in renewable energy production, including wind (we'll look at quite a few examples of municipal wind projects shortly).[12]

In Europe, the experience of community wind in Denmark and Germany is particularly instructive. Not only are the wind turbines, or clusters of turbines, distributed across the landscape, the ownership is also spread across hundreds of thousands of individual participants as well. One-fourth of the wind generating capacity in Denmark, amounting to an equivalent of €1 billion, has been developed by windmill guilds (*vindmølleaug*) roughly equivalent to what would be called cooperatives in North America. An additional sixty-five percent of capacity has been installed by Danish farmers. And in Germany, as much as one-third of the nation's wind capacity has been built by associations of local landowners and residents (*Bürgerbeteiligung*). Individual German investors have installed as much as 9,000 megawatts of wind-generating capacity amounting to an investment of over €10 billion ($13.2 billion). More than 100,000 households (or nearly 5 percent of the population) in Denmark, and another 200,000 people in Germany own a share of a nearby wind turbine or shares of one in their community.[13]

DENMARK: A COOPERATIVE EFFORT

Although a number of different mechanisms to develop community wind projects have been used in Europe, they all require the participants to work together to achieve their common goal. This strategy is regularly employed in northern Europe for a wide variety of purposes, but especially in Denmark. Cooperative wind development has been a natural outgrowth of Danish cultural and agricultural interests. Around 1980, the Danish parliament provided incentives for wind cooperatives to encourage individual action toward meeting the nation's energy and environmental policies. This program enabled virtually any household to help generate electricity with wind energy without necessitating the installation of a wind turbine in their own backyard.[14] There were three key components to the Danish wind initiative:

1. The right of wind power developers to connect to the electrical grid
2. The legal requirement that utilities purchase the wind-powered electricity, and
3. A guaranteed fair price.[15]

As part of the legal requirements, the wind generator pays for the cost of connection to the nearest acceptable point on the grid, while the grid operator pays for any additional expenses required for grid upgrades and other improvements. These requirements removed one of the biggest hurdles to developing the wind industry in Denmark. As a result, the entire wind-power initiative was spectacularly successful, and the idea caught on in Germany and the Netherlands as well.

Here's how the strategy worked. Danish law encouraged mutual ownership of wind turbines by exempting the owners from taxes on the portion of the wind generation that offset a household's domestic electricity consumption. A wind co-op would then purchase a wind turbine, select the best site available, sell electricity to the electric utility under favorable terms, and share the revenues among its members. This enabled the group to buy the most cost-effective turbine available, even though it may have generated far more electricity than individual co-op members needed for themselves. Although, technically speaking, many of these cooperative

Lynetten cooperative turbines in Copenhagen, Denmark. Paul Gipe.

ventures were set up as limited liability companies owing to the vagaries of Danish law and tax policies, the cooperative nature of the organizations was clear. The Lynetten cooperative (Lynetten Vindkraft I/S) that owns four out of the seven prominent, 600-kilowatt wind turbines on a break-water within Copenhagen's port is a good example of the success of the cooperative wind strategy in Denmark.[16] With over nine hundred members, the cooperative is one of the larger wind co-ops in the world.

Middelgrunden

But perhaps the most famous Danish cooperative of all is Middelgrunden, which in March 2001 was the world's largest offshore wind farm and is still one of the largest that is cooperatively owned. Located on a shoal about 2 kilometers outside of Copenhagen harbor, the Middelgrunden wind farm consists of twenty Bonus Energy 2-megawatt wind turbines with a hub height of 64 meters (210 feet) and a rotor diameter of 76 meters (250 feet). The twenty turbines are installed in a gentle arc with a total wind farm length of 3.4 kilometers (2 miles) and a combined generating capacity of 40 megawatts. Ten of the turbines are owned by the Middelgrunden Wind Cooperative, while the remaining ten are owned by Copenhagen Energy, the local municipal electric company. The joint relationship between the coop-erative and the electric company proved to be extremely helpful throughout the planning, approval, and construction phases of the project, and provides an excellent model for North American wind initiatives as well.

The Bonus 2-megawatt units are specially designed for the harsh mar-itime climate in which they operate with high-grade external paint; her-metically sealed machinery; and a self-contained, internal climate-control system. The surface of the fiberglass blades is similar to fiberglass boat hulls and consequently requires no additional corrosion protection.[17] A built-in crane is used for turbine maintenance. The status of the turbines and their output is monitored and regulated by an advanced, computerized control system, and the power from the wind farm is carried via an underwater cable to the mainland.[18] Performance of the turbines is also publicly posted on the co-op's Web site (www.middelgrund.com). The turbines were pre-assembled on shore in three parts before being floated to the concrete foun-dations (which were also built on shore and towed into place) and erected by a barge-mounted crane.

Middelgrunden wind farm. Hans Chr. Soerensen.

The original idea for the wind farm was conceived in 1993. The partnership was founded in May 1997 by the Working Group for Wind Turbines on Middelgrunden, with the goal of establishing and managing a wind farm on the Middelgrunden shoal. Partners own a share in the venture corresponding to 1/40,500 of the partnership per share purchased.[19] Organizers believe that the project would not have been built without the public support generated by early public involvement in the planning process and local ownership. Despite numerous obstacles, the cooperative sold 40,500 shares for €570 ($678) each, a price set as low as possible to encourage broad participation. The total investment budget for the co-op was €23 million ($27 million). All together, about 8,500 investors bought shares.[20] The projected payback time on the investment was estimated at eight years, with a rate of return of 7.5 percent after depreciation. Despite some initial technical problems, overall the wind farm has performed well and met or exceeded electrical-generation projections, providing enough electricity for more than 40,000 households in Copenhagen.

Samsø

Another excellent example of community-owned wind is Samsø. During 2002, a cooperatively owned offshore wind farm was installed near the small Danish island of Samsø. In 1997, the island (with a population of 4,400) was selected in a national competition to be a demonstration of a community that would be totally supplied by renewable energy within ten years. The wind farm is one of a wide range of renewable energy strategies that are part of the Renewable Energy Island project, including district-heating systems based on straw, wood, and solar energy, as well as individual stoves, heat pumps, and solar collectors for houses outside of villages. The offshore wind farm, consisting of ten Bonus Energy 2.3-megawatt turbines, was developed by a cooperative with local people and municipalities as members. About 440 of the 2,000 households on the island own shares in the wind turbines. The offshore project complemented an existing eleven-turbine, onshore wind farm, making the island completely self-sufficient in its electricity production. What is especially remarkable about the Samsø project is that it managed to maintain momentum even after a new conservative Danish government was elected in 2001, which sharply reduced subsidies for renewable energy, as well as support for organizations like the Energy and Environment Office. Søren Hermansen, the office manager, attributes this success to the broad local backing the initiative has achieved. "People here regard this project as vital for the future of the whole community," he says.[21] The Samsø turbines have performed extremely well and are among the most productive wind farms in the world. There are six offshore wind farms in Denmark. It's interesting to note that at least one other offshore project that was not community-owned experienced delays due to strong local opposition.

A Postscript

Denmark has been rightly viewed as one of the best examples of what can be accomplished with renewable energy when a government follows enlightened, forward-looking policy that includes its citizens as partners. Unfortunately, this is no longer the case. After the original Danish wind power subsidies were enacted in 1980, the wind industry took off and made great progress. A change in government in 1988 brought the Neoliberal-Conservatives into power, who promptly cut the subsidy in half. Nevertheless, the basic core of the original initiative remained in place, and in 1993 the Social Democrats returned to power and remained in office under various coalitions until 2001. This is now viewed in retrospect as the

THE NETHERLANDS: KENNEMERWIND COOPERATIVE

Denmark is not the only country with numerous wind cooperatives. Many of the first wind turbines installed in the Netherlands were installed by co-ops, although conditions were far less favorable than in Denmark. One of the largest is Coöperatieve Windenergie Vereniging Kennemerwind. The cooperative operates ten turbines in all, nine of which are part of a group of fifteen turbines along a canal in Noord Holland.

Kennemerwind has no bank loans or debt. It raises all of its capital from members, most of whom invest as little as €50, though some have invested as much as €10,000 to €15,000. The co-op views its members' investment as a fifteen-year loan, which is to be repaid. While the driving force behind the coop's 650 members is their desire to produce clean energy, Kennemerwind consistently pays an annual dividend of 7 percent to its shareholders. Members can reinvest their dividends, and nearly all do.

From 1989 through 2002, the coop generated more than 15 million kilowatt-hours. With the addition of several new Lagerwey turbines in the mid-1990s, Kennemerwind began producing from 1.5 million to nearly 2 million kilowatt-hours yearly, enough electricity to meet the needs of 500 to 650 Dutch households. During the life of the co-op, the wind turbines have earned more than a million Euros.[22]

Kennemerwind. The four Lagerwey turbines (right) along the Noord Holland Kanaal are part of the Kennemerwind co-op. Each turbine generates 150,000 to 200,000 kWh per year. Paul Gipe.

golden age of wind power in Denmark, during which production tripled and 85 percent of the approximately 5,300 turbines in the nation were owned by local cooperatives and farmers. Then in 2001 the Neoliberal-Conservative government was reelected and almost immediately pulled the plug (à la Ronald Reagan) on the wind industry, saying that it had to stand on its own in the "free" market (the same "free" market that subsidizes fossil fuels and the nuclear industry around the world to the tune of hundreds of billions of dollars annually). The small, locally owned wind co-ops, which had been the very foundation of the industry's success in Denmark, suddenly found themselves marginalized and discriminated against in favor of large corporate projects. In 2004, only five wind turbines were erected in Denmark, and the industry was in severe decline.[23] "Denmark was a shining example of how to build a world-leading industry," says Paul Gipe. "Today, it's an example of how to destroy that industry almost overnight."[24] The cost of success, it seems, is constant vigilance, especially when it's a matter of the people versus large corporate interests.

GERMANY

With more than 18,000 megawatts of installed capacity by the end of 2005, Germany is by far the world leader in wind power, according to Paul Gipe. What's more, a leading industry consultant says that Germany's installed wind capacity is expected to surpass 26,000 megawatts by the end of 2009.[25] Despite the fact that the nation's wind resources are not the best in Europe, Germany owes much of its success in wind generation to its favorable *Stromeinspeisungsgesetz*, or electricity feed-in law (EFL), implemented in 1991. The EFL provided a stable and profitable foundation for wind projects and propelled Germany to the forefront of installed capacity. In addition, the EFL encouraged the development of a strong wind-turbine manufacturing base. The feed-in law has also encouraged community wind projects by requiring German utilities to pay wind generators 90 percent of the average retail electricity price. This has encouraged significant wind-farm development, especially in the northern coastal regions of the country that have the best wind resources. I have traveled extensively through this region, and can say from personal experience that the wind farms there are simply amazing, and that they have not in any way hurt tourism or spoiled the landscape. You quickly get used to them.

In 2000, the EFL was replaced with the *Erneuerbare-Energien-Gesetz* (EEG), or Renewable Energy Act. The EEG is an update of the previous law that essentially brings the feed-in tariffs[26] in compliance with current European Union law. An amendment to the EEG, in the summer of 2004, provides a fixed reference point for wind-farm operators; for every kilowatt-hour of power fed into the grid from a wind turbine erected in 2005, they receive 8.53 Euro cents ($.10) from the local grid operator for the first five years and 5.39 Euro cents ($.06) thereafter, according to the German Wind Energy Association. This minimum price allows the wind-power sector to compete effectively against the coal and atomic energy industries. As in Denmark, the wind generator pays for the cost of connection to the nearest acceptable point on the grid, while the grid operator pays for any additional expenses required for grid upgrades and other improvements.[27]

The most common forms of legal structure for community wind projects in Germany are the German equivalents of limited liability companies or invest-ment co-ops. These associations can be limited to landowners participating in the project, or the project may be opened to investors from surrounding com-munities, or sometimes nationwide. Since these legal structures are not directly linked to individual investors' electricity consumption (as they often are in Sweden and Denmark), many wind projects in Germany tend to be larger, with higher minimum investment thresholds.[28]

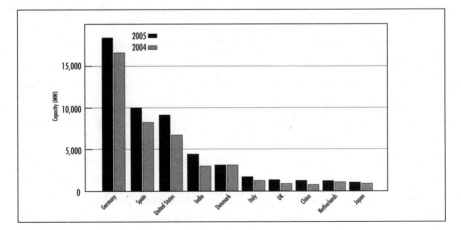

Total installed windpower capacity. World Wind Energy Association.

There are many examples of cooperatively owned wind farms in Germany. One relatively early example is Windenergiepark Udenhausen-Mariendorf, a cluster of five Vestas V44s wind turbines on 53-meter- (174-foot-) tall towers installed in April 1996. The wind farm is located in the Mittelgebirge in Germany's central highlands just northeast of the city of Kassel. The turbines are sited in the middle of a rapeseed (canola) field on an intermediate ridge within a broad valley bounded by the higher Mittelgebirge. The wind farm generates about 5 million kWh annually. The cooperative has sixty-five shareholders, and shares of the project were sold for 2,500 DM each ($1,500). The share price was deliberately kept low so more local people could participate. Most of the investors live in the surrounding villages, though some live in the nearby city of Kassel. The shares provided 30 percent of the total investment. The Deutsches Ausgleichsbank, through a local bank, provided a loan for 45 percent of the project, while the state of Hesse provided a 25 percent capital investment subsidy. The loan for the project will be paid off in June 2007. Unlike other similar projects where the land is rented, the land for this project was purchased from the farmer.[29]

With over 16,826 wind turbines installed by 2005, generating over 6 percent of the country's electricity, the German wind industry is well positioned for continued growth. The change in national government after the elections in September 2005 may eventually signal a shift in federal policy regarding wind power, but early signs are reassuring. Nevertheless, despite a trend toward larger and larger projects, about 30 percent of installed wind turbines in Germany are still owned by the public, mostly through associations of landowners and local residents, and community-owned wind is expected to do well in the market for the foreseeable future.

SWEDEN

Sweden was one of the early pioneers in wind power, with a development program that began in 1975. At first, the industry grew very slowly, but after the creation of a new national energy plan and introduction of an investment subsidy in 1991, the number of turbines owned by small private enterprises, partnerships, cooperatives, and other nonutility groups increased considerably. After a decision to close the Barsebäck II nuclear power plant in 2005, the Swedish government agreed to speed up the installation of several large offshore wind farms. The government also announced plans to encourage further development of wind power, both onshore and offshore, with a program to streamline the approval process, and encourage greater cooperation between turbine manufacturers, public authorities, and other stakeholders at the central, regional, and local levels. The long-term goal is to produce 100 percent of the nation's energy through renewable resources.

A culture of cooperation is encouraged in Sweden's educational system, and this attitude permeates much of society. Consequently, cooperatives are a familiar model, especially at the local level and in the agricultural sector. Sweden, like Denmark, has had government policies in place for over twenty-five years that supported the steady development and growth of a wide range of renewable energy strategies. These initiatives have also supported the involvement of cooperatives. Energy taxes and favorable electric feed-in tariffs have been instrumental to the success of the country's renewable energy programs, and also produced the revenues that have supported many projects.

Sweden's energy generation systems are composed of a large number of small power stations scattered across the nation. This distributed-generation pattern has created a system that offers many benefits to local communities, which have come to view these facilities favorably, since they provide opportunities for local cooperative ownership and employment. Local planning authorities also support these projects, since they view them as contributors to the local economy. Over fifteen thousand households in Sweden are investors in local, cooperative wind projects, although the momentum for these projects has slowed since 2000, with a shift in government policy toward a more "free market" approach to energy policy.

Community wind-ownership strategies have existed in Sweden since 1989, when the founders of Vindkompaniet AB started the first Vindsamfällighet (wind real estate association) on the island of Gotland, where much of Sweden's installed wind capacity is located. Gotland, with 58,000 inhabitants, has plans to be fossil-fuel-free by 2025. More than two thousand households on the island own shares in wind turbines through local wind energy co-ops, generating returns of up to 7 percent annually.

The real estate association is a legal structure somewhat peculiar to Sweden and not easily transferable. The other popular ownership model, however, is the *Vindkonsumföreningar* (wind consumer cooperatives). This model is similar to consumer cooperatives in North America, and could be adapted for use by wind co-ops here. Regardless of the legal distinctions between these models, however, the community ownership concept is central. Up until fairly recently, these Swedish communes and cooperatives have operated locally, selling their electricity to the local utility at an agreed upon feed-in tariff.[30]

The Glimminge Wind Power Real-Estate Association is located in the coastal town of Höganäs in southern Sweden. The commune consists of fifteen farmers and four other individuals, all local residents, who communally own a 500-kilowatt Wind World turbine located in one of the farmer's fields. The project, which was completed in early 1995, received a 35 percent capital investment subsidy from the government, with the remaining 65 percent raised through the sale of 950 shares of 1,000 kWh/year. Each share cost SEK2,500 (about $320), and the number of shares per owner varies from 10 to 250, depending mainly on each owner's energy consumption. The two farmer/promoters say that they were not motivated by environmental concerns, but instead viewed the investment solely as a good business opportunity.[31] There are many other cooperatively owned wind farms scattered across the country.

With the drop in wind feed-in tariffs following the opening of Swedish residential electrical markets to competition in 1999, a cooperative and a distribution utility have created an innovative partnership that may help to overcome many of the investment and membership constraints that have faced local cooperatives in the nation. The distribution utility, Falkenberg Energi, agreed to provide service to Sveriges Vindkraftkooperativ (Sweden Wind Cooperative) on a nationwide basis, allowing the cooperative to sell wind power directly to members throughout the country. This agreement in effect created Sweden's first nationwide cooperative, and industry observers expect this model to increase significantly the number of cooperative wind investors in the country in the coming years.[32]

UNITED KINGDOM

The United Kingdom has the best wind resources in Europe. It has been estimated that just over 1 percent of the total land area in the United Kingdom could supply about 20 percent of the total electrical needs from wind. However, the United Kingdom only ranks fifth in terms of installed wind-generation capacity. This is due in part to muddled government policy that generated strong local opposition to the visual impact of wind turbines on the landscape. A lack of experience and knowledge, along with the difficulty of obtaining planning permission, have been the main barriers to community-owned wind farms. But that's beginning to change with a small but growing number of co-op wind projects being proposed and approved. The key to the success of these proposals is that they are locally inspired and locally owned.

Baywind Energy Cooperative

Baywind Energy Cooperative Ltd. (www.baywind.co.uk), located at Harlock Hill near Ulverston in Cumbria, is the nation's best example of a cooperative wind project. Located on an open stretch of windswept farmland overlooking Morecambe Bay, the 2.5-megawatt wind farm consists of five Wind World turbines, each rated at 500 kilowatts. These machines produce enough electricity to meet the average needs of 1,300 typical homes. Established in 1996 by a pioneering Swedish company as the United Kingdom's first community-owned renewable energy cooperative, Baywind got off to a shaky start when it ran into approval and regulatory problems, and the Swedish company ran out of money. The Swedes went home, but the struggling co-op managed to recover from its initial setbacks with the help of a loan from the Co-operative Bank, and subsequently raised £2 million through share offers with over 1,300 members. Baywind also now owns one of four turbines at Haverigg II, located on an abandoned airfield on low-lying coastal land near Millom in Cumbria. Including Haverigg II, Baywind now owns a total of six wind turbines and is actively working on additional projects elsewhere. About 40 percent of Baywind shareholders live either in Cumbria or Lancaster, with a larger number from the Northwest Region. Baywind has attracted a lot of interest from many other communities in the United Kingdom who would like to replicate its success.[33]

Energy4All

In 2002, Baywind also established Energy4All to support and encourage the ownership and operation of renewable energy projects by local or community-based co-ops. This project development company is owned by the co-ops it creates, forming a commonwealth of cooperative enterprises. In Scotland, a number of small Highland communities are being assisted by the Highlands and Islands Enterprise to establish community-owned wind farms to provide both a high degree of self-sufficiency in power and a source of steady income. Energy4All has been advising these communities on suitable financial structures to support these initiatives.

In addition, Energy4All has developed an innovative new model for local participation in larger projects. The company has signed a number of unique agreements with carefully selected developers to enable communities to establish local co-ops in conjunction with major new wind-farm sites being planned in Scotland, England, and Wales. This allows the community to acquire a stake in the project, and the opportunity to develop associated activities without the high risks and costs of developing the entire project as a community-owned enterprise. The first of these projects was launched in early 2006 in Aberdeenshire; others will follow over the next few years.[34]

Most significantly, Energy4All has recently launched its first major new 100 percent community-owned project, at Westmill on the Oxford-Wiltshire border in the southeast of England. This site had been developed over twelve years by an organic farmer who wanted to see community-owned renewable energy generated on his land. Energy4All has taken over the project and launched a national share offer for the Westmill Wind Farm Co-operative in November 2005. This offer raised £3.75 million ($6.53 million) to create the largest community energy project in the United Kingdom. Each share is worth £1 ($1.74), with a minimum investment of £250 ($436) and a maximum of £20,000 ($34,876). Construction on the 6.5-megawatt wind farm, originally slated to begin in the summer of 2006, has been delayed due to extremely high global demand in the wind turbine market. The five 1.3-megawatt wind turbines will eventually provide enough electricity to power 2,500 average homes.[35] Community-owned wind is expected to grow substantially in the United Kingdom in the coming years.

UNITED STATES

David Blittersdorf, the cofounder and current chief technology officer of NRG Systems of Hinesburg, Vermont, agrees with Paul Gipe about the potential for cooperative or community-owned wind initiatives in North America. "The potential is huge," he says. "Denmark and other European countries have had agreements for many years for the co-ownership of wind turbines and for feeding the power generated by them to the grid, but in the United States we're a bit behind on how to do that. The main challenge is getting our rules and regulations right so the cooperative concept can be implemented here."[36] NRG Systems is a global leader in wind assessment technology, and Blittersdorf is a past president of the American Wind Energy Association. Blittersdorf and others in the wind industry have been working tirelessly to remove the many barriers to cooperative wind in North America.

Group Net Metering

One of the biggest barriers is connecting to the grid—and getting paid a fair price. One key strategy with a lot of promise that addresses these issues is called *group net metering*. About forty states and the District of Columbia currently have some form of net-metering laws on the books. The state of Vermont, for example, has had a net-metering law since 1998. The law has been amended several times since, and now allows any electric customer in the state to connect a wind, solar, or fuel-cell electrical system to the local utility after receiving a "Certificate of Public Good" from the Vermont Public Service Board. The 2002 amendments to the law include a provision for "group net metering" for farmers with methane-electric generation systems and multiple electric meters.[37] Renewable energy advocates in the state have been pushing hard to expand the law to include photovoltaic and wind-powered systems, as well as raising the cap on system output to higher levels, offering a wider range of possible models for group ownership of these initiatives. Unfortunately, despite recent incremental improvements, the law is still far too restrictive to be genuinely useful, especially for community-supported energy projects.

The concept is fairly simple. The electrical output from a group-owned wind turbine (or turbines) is metered at the point where it connects to the grid. Then the owners split the production credit based on their percentage of ownership in the turbine, and the amount is credited to their home electric bill by the utility. "The only limitation would be that this can't take place across utility boundaries," Blittersdorf explains. "So as long as all the members are served by the same utility, it would be really easy and just a matter of allocating their respective outputs on their electricity bills at retail rates."

Favorable legislation is important, according to Blittersdorf. "Net metering has been pushed in the past five years or so across the United States, and now, with the federal Energy Policy Act of 2005 stipulating that net metering will soon be a national law, it's just a question of getting the rules right. I see a changing world—there's starting to be some real momentum for this—and the utilities are becoming a little more cooperative because they see a changing world too."[38]

Advantages

A cooperative or community-owned wind project offers many advantages. It stimulates the local economy by creating new jobs and new business opportunities for the community while simultaneously expanding the tax base and generating new income for local residents. Locally owned wind also generates support from the community by getting people directly involved. Another major advantage of a community-owned project over an individually owned turbine is location flexibility. Most individual systems are installed on the owner's property, which may or may not be an ideal wind site. A co-op turbine, on the other hand, can be located wherever the best site can be found within the service area of the group's utility. This advantage alone should be sufficient justification for pursuing the co-op model. But community wind projects offer yet another advantage: they retain a greater amount of income in the local area and increase the economic benefits substantially over projects owned by out-of-area corporate developers, according to a study conducted by the National Renewable Energy Laboratory for the Governmental Accountability Office. NREL compared the effect of a 40-megawatt corporate wind farm owned out of area and twenty 2-megawatt wind plants owned locally. Looking at eleven locations, the study found local ownership yields an average of $4 million in local income annually, over three times more than the $1.3 million produced with out-of-area control, while job creation was more than twice as large in the local model.[39]

Blittersdorf is optimistic about community supported wind and its many benefits, and he offers some advice for would-be local wind developers. "The first thing is to figure out the technical aspects and identify a proposed site," Blittersdorf advises. "Then bring in the folks in the wind industry who are knowledgeable about equipment and sources and all the other details so that you have very little technical risk going into the project. After you have gotten your finances in order, most of your remaining hurdles are governmental and regulatory issues."

"It's really a matter of getting a model in place like the German feed-in law that stimulated their wind and solar industries," he continues. "It's amazing what can be accomplished when you get the right regulations and laws in place, then you can really drive a market. That is what has to occur here in order to get over the tipping point. Once that has been accomplished, having some folks in the local community that are energetic and interested in developing the cooperative concept and who are willing to make it happen is what is needed. In the end, that's going to benefit both the community and the local investors because they are hedging their future energy supply and everyone can win."[40]

Lisa Daniels, the executive director and founder of Windustry, a Minnesota-based nonprofit organization working to create an understanding of wind-energy opportunities for rural communities, is also enthusiastic about the potential for locally owned wind projects. "In this country, most wind turbines have been set up by developers from outside of the community," she notes. "So most of the time rural landowners are leasing their land to external wind developers and they are just receiving an annual royalty. But those wind projects start to look a lot better if they are in smaller clusters, and when there is local ownership of those clusters, which keeps much more money circulating in the local economy. I think there will be greater acceptance of the changes to the landscape caused by these projects if they are smaller and if they are locally owned. I also think the wind industry in general, and community wind in particular, is on the threshold of dramatic growth."[41]

Production Tax Credit

In the United States, federal support for the wind industry has been mainly through what is known as the energy production tax credit (PTC), passed as part of the 1992 Energy Policy Act. Originally designed to support electricity generated from wind and certain bioenergy resources, the credit provides a 1.8-cents-per-kilowatt-hour benefit for the first ten years of a facility's operation. In theory, this sounds like a useful tool for the wind industry. In practice, it's been problematic, since the PTC has been at the mercy of the political process, and has been allowed to expire on three separate occasions. These on-again, off-again politically motivated gyrations have created a rollercoaster, boom-bust cycle that has plagued the wind industry in the United States. But the larger problem for small landowners and community-based wind projects is that they simply don't have large enough tax liabilities to benefit from the PTC in the first place. A number of states across the country have begun to step in to overcome some of the many hurdles to local ownership of wind projects, including Minnesota, Iowa, Illinois, Oregon, Washington, New York, Colorado, Massachusetts, and Ohio. Others are expected to follow soon.

Minnesota

Although the benefits of community-owned wind are numerous, examples of successful projects in the United States are few and far between, but their numbers are beginning to grow. Of all the states in the nation, Minnesota has unquestionably led the way in promoting and supporting locally owned wind projects, and some of the best examples are located there. In 1997, and again in 2003, the state offered per kilowatt-hour production incentives, which function much like European feed-in laws. In addition, a number of creative ownership structures—sometimes referred to as the "Minnesota flip" owing to their unusual ownership-swapping strategies—have been developed that allow small wind projects to take advantage of production tax credit benefits.[42] "Farmers and small businesspeople really can't use the PTC on a utility-scale turbine," explains Dan Juhl, the developer of the "flip" strategy. "The flip helps us to attract outside equity investors with an appetite for the tax credits into a community project. They come in on the front end with equity, then they extract the tax values out of the deal, and after ten years the tax credits are gone and the owner-ship "flips" or reverts to the farmer or local community who end up owning the wind machine or wind farm."[43] The state has implemented other poli-cies as well, including standardized purchase and interconnection agree-ments, and a state renewable energy standard that requires a 160-megawatt share of production capacity be reserved for smaller, locally owned projects under 2 megawatts.[44] Combined, these policies are expected to spur 460 megawatts of locally owned wind turbines in the state.[45] But that's not all.

Passed into law by the 2005 Minnesota legislature, the Community-Based Energy Development (C-BED) initiative replaced the ten-year production incentive of $0.015/kWh by establishing a framework for qualifying owners of wind generation projects to negotiate power-purchase agreements (PPAs) with all Minnesota electric utilities. The negotiating framework sets a price for electricity that is based on the net present value of energy over a twenty-year PPA, significantly stabilizing the long-term viability for community-based wind-power production in the state. The C-BED initiative provides numerous benefits—including very competitive, declining-cost renewable energy over the twenty-year life of C-BED projects—while also producing viable cash flows for project owners and developers. As a result of the C-BED initiative, residents, local business associations, local and tribal units of government, school districts, institutions of higher education, and ordi-nary people who live in and consume electricity in Minnesota now have a better opportunity to develop C-BED projects.[46] "Keeping our energy dollars in the local community is a huge economic tool, and the C-BED program has brought that to the attention of our policymakers and legislators," says Dan Juhl, who has played a key role in the C-BED program. "Minnesota has enacted a statute that says we want renewable energy as a preference. So, if we are going to do that anyway, let's leverage it into something even greater—local economic development. When that happens, we get clean, sustainable energy and economic development which helps stabilize rural communities."[47]

All this activity has provided fertile ground for community wind in Minnesota, and a variety of local projects have been sprouting up in fields around the state in recent years. These projects include RiverWinds, a six-turbine municipal utility project in Worthington; a voluntary "Capture the Wind" initiative by Moorhead Public Service in Moorhead using two turbines; single-turbine public school installations in Madison, Plymouth, and Pipestone; Carleton College in Northfield; Macalester College in St. Paul; and the University of Minnesota, Morris; as well as a number of local and farmer-owned initiatives. One of the best examples of a locally owned wind development, however, is MinWind.

In 2000, a group of farmers in Luverne, Minnesota, began to develop a plan for farmer-owned wind turbines that would take advantage of state policies favoring local wind development. Their goal was to find an investment that would generate new income for farmers and provide economic benefits for the local community. The rapid growth of the wind industry around the country and the success of wind farming on the nearby Buffalo Ridge made wind energy a natural choice. "We wanted a farmer-owned project that would bring economic development, get farmers a return on their investment, and could use local businesses and contractors to do the work," said Mark Willers, a project leader and farmer from Beaver Creek, Minnesota.[48]

To develop their idea, the group conducted extensive research and finally settled on forming two limited liability companies (LLCs), Minwind I and Minwind II. This legal structure was the best option because it maximized the companies' ability to use tax credits and other incentives for wind energy while maintaining some principles of cooperatives such as voluntary and open membership, democratic member control, and concern for the greater community.

Of the total $3.5 million cost of the project, 70 percent was financed by a local bank. The remaining 30 percent was equity-financed. Sixty-six investors from the region surprised many observers when they quickly snapped up all the available stock (at $5,000 per share) for the $1.1 million equity share in both companies in only twelve days. All of the members are from Minnesota. Rules state that 85 percent of the shares must be owned by farmers, leaving the rest for local townspeople and nonfarmers, who could someday inherit shares. Each share gives the owner one vote in the company, and no single person can own more than 15 percent of the shares.

The two companies were formed to take advantage of a Minnesota renewable production incentive that provides a 1.5-cents-per-kilowatt-hour payment for wind projects up to 2 megawatts for the first ten years of production. Although they coordinate closely, the two companies are governed by separate boards of directors, have different groups of investors, and maintain separate financial records. Both groups also relied heavily on expertise from consultants to develop the actual wind project and negotiate the power-purchase agreement, and on a team of lawyers to determine the business structure.

Each project consists of two NEG Micon 950-kilowatt turbines, and all four turbines are located on the same farm seven miles southwest of Luverne. The site was chosen because the group wanted to use land owned by one of the project's investors, and this particular farm had the best combination of wind resource and access to transmission lines. According to Willers, the biggest challenge was not the technology, the wind resource, or raising the money. The biggest obstacle, rather, was negotiating a power-purchase agreement with a utility that would buy the wind-generated electricity at a fair price. Discussions with the local rural electric cooperative did not work out owing to a wide range of issues. Eventually, after months of negotiation, the wind-farm promoters entered into a fifteen-year contract with Alliant Energy, which uses the power to help satisfy renewable energy standards in other states. "The key reason why MinWind was such a success was that they were able to organize their community around the power contract, and bring their community investment into it to make it work," says Dan Juhl, who played a significant role in the project. "They did a great job at that, and it's a great model for other people to follow."[49]

The MinWind I and II projects were so successful that the organizers decided to pursue opportunities for additional projects. After a lot of hard work, MinWind Energy dedicated MinWind III-IX, seven new 1.65-megawatt NEG Micon wind turbines on December 3, 2004. These turbines are owned by approximately two hundred local investors, following the same principles as the original MinWind I and II projects. Willers and many others have invested countless hours in developing the MinWind projects, but they believe their efforts have been worthwhile. "We've spent an incredible amount of time on this, but we needed to do it for our community and our friends who are farmers," says Willers.[50]

Massachusetts

In the state of Massachusetts the concept of "community wind" has taken a slightly different form. Hull Municipal Light has been serving electricity consumers in the town of Hull, Massachusetts, since 1894. In December 2001, the state's first modern wind turbine was installed on Windmill Point at the tip of the town of Hull. The 660-kilowatt Vestas V47 turbine, with a hub height of 50 meters (164 feet) and rotor diameter of 47 meters (154 feet), is owned and operated by the Hull Municipal Light Plant (HMLP). The project was developed with the assistance of the University of Massachusetts Renewable Energy Research Laboratory and the Massachusetts Division of Energy Resources. The turbine is located close to the local high school, within 100 feet of the site of a 40-kilowatt Enertech turbine (since removed) which was installed in the 1980s. The town is in a highly populated coastal area, within eight miles of Boston's city hall, and only five miles from the busy runways at Boston's Logan International Airport.[51] At first blush, this might seem to be an unpromising location. Nevertheless, Hull Wind I, as the project is now called, had a lot going for it from the beginning.

There were a number of factors that led to the project's success. First and foremost was the local nature of the initiative. The HMLP was an active and enthusiastic participant from the beginning. The project also had strong support from a number of local champions who were able to generate enthusiasm and momentum at the grassroots level. The town had previous experience with wind (the 40-kilowatt turbine), and local residents also realized that there would be a public benefit for the whole community.

Hull Wind I. Alex Mavradis Photography.

NOT IN MY BACK YARD (NIMBY)

With the proliferation of wind farms and wind proposals in the United States, there has been a rising tide of opposition from some local residents who believe that they will be negatively affected, especially by large-scale commercial projects. In general, wind opponents represent a small, but vocal minority, who object to a number of issues, but especially to the visual impact of the projects. Wind opponents are also concerned about the effects of the noise created by the turbines on wildlife, the danger posed by the spinning blades to birds, as well as the impact of service roads and transmission lines associated with the projects. While some of these opponents reject wind altogether, others claim to be generally in favor of wind—but not in their back yard. These issues have also divided the environmental community.

No project better showcases these issues than Cape Wind in Nantucket Sound. The Cape Wind project—the first major offshore wind farm in the United States—is planned for construction on Horseshoe Shoal, five miles off the Cape Cod shore in Massachusetts. The wind farm will consist of 130 wind turbines, with a total maximum output of 420 megawatts. In average conditions the project will produce enough electricity to power three-quarters of the Cape and nearby islands.[52]

Residents of the Cape are divided. Although many residents are in favor of the project, others, most notably former Governor Mitt Romney and historian David McCullough, as well as Robert F. Kennedy, Jr., a senior attorney at the National Resources Defense Council (NRDC), have vocally opposed the wind farm. They have a lot of support among members of the Alliance to Protect Nantucket Sound (funded in large part by owners of property overlooking the sound),[53] which also opposes the project. In a January 6, 2006, op-ed piece in the *Cape Codder*, Kennedy said that he supports wind power, but not in Nantucket Sound. He cites many of the issues mentioned above, as well as Humane Society estimates of thousands of bird kills every year and estimates from Boston's

Suffolk University of up to 2,533 job losses due to predicted declines in tourism. Kennedy also contends that the project's underwater cables would interfere with the local fishing industry.

Many environmentalists and organizations—including Greenpeace, the Union of Concerned Scientists, World Wildlife Fund, and even the NRDC itself—disagree with Kennedy's stand. Many other supporters of the project disagree as well. They refute many of Kennedy's claims about the potential environmental hazards and noise pollution, and further argue that the project would have a minimal effect on the fishing industry, since the cables would be buried 6 feet beneath the seabed.[54] Even the visual impact of the project is questionable, since, on a clear day, the slender towers would only appear as tiny, half-inch figures on the distant horizon. And contrary to the claimed damage to the tourist industry, project supporters predict that just the opposite will occur, since similar offshore wind farms in Europe have proven to be big tourist draws.

Bill McKibben, the well-known environmentalist and author of *The End of Nature* (among many other books), offers his perspective on the debate. "The environmental movement is reaching an important point of division between those who truly get global warming, and those who don't," he says. "By get, I mean understanding that the question is of transcending urgency, that it represents the one overarching global civilizational challenge that humans have ever faced. That it's as big as the Bomb."[55] McKibben is not the only one to view the issue in these stark terms. The debate comes down to weighing local NIMBY concerns against global climate concerns, according to John Passacantando, executive director of Greenpeace USA. "I respect people who wage NIMBY battles—the environmental movement was founded on people protecting their local, sacred areas," he says. "But today, solving the climate crisis has become so urgent that it trumps NIMBYism. It's as simple as that."[56]

(HMLP used the "profits" from the wind turbine to eliminate billing for the town's street lights.) There was strong technical support throughout the process. Last, but by no means least, the project had an available site with great wind resources.

Over the years, Hull Wind I has performed exceptionally well, and in just over three years of operation, reached the "break-even" milestone of 5 million kilowatt-hours. By the end of 2005, the turbine had generated over 6 million kilowatt-hours of electricity.[57] Over the twenty-year lifespan of the project, the turbine is expected to save the town about $3 million. This award-winning project has been so successful the town decided to add a second turbine in 2006. The new project, Hull Wind II, is a Vestas V80 1.8-megawatt turbine sited at the town dump, a capped landfill. The new turbine is projected to have about three times the output of Hull Wind I.[58] The 330-foot, $3 million turbine arrived in January 2006, and was installed and operating by May. "The . . . thing about Hull is it seems they move faster down there than anybody else," says Frank Gorke, a consumer advocate and energy specialist at the Massachusetts Public Interest Group. "They're already . . . build[ing] a second turbine before anybody else has a first."[59]

There is another promising community wind initiative in Massachusetts, the Community Wind Collaborative (CWC). Launched in September 2003 by the Massachusetts Technology Collaborative through its Renewable Energy Trust, the $4 million initiative was designed to help Massachusetts communities evaluate, design, construct, and operate smaller wind projects (4.5 megawatts or less) to both reduce energy costs and contribute to a cleaner environment. One of the biggest challenges for this strategy is that projects of this size and complexity are difficult for individual towns to tackle on their own. Yet, at the same time, they are too small to be of interest for most large developers. The idea for the collaborative grew out of the sharp differences between the highly charged debate about the large, 420-megawatt, offshore Cape Wind project and the well-received municipal Hull Wind I project. The collaborative hopes to fill the gap between these two models, by encouraging mid-sized projects that local communities will embrace and that also will benefit the residents involved. Around forty communities have expressed interest, and are at various stages of project development.[60]

The town of Orleans on Cape Cod is the furthest along in the process.

The preliminary plan calls for two Vestas 1.5-megawatt turbines that would power the town's new water treatment plant. Although the project still has significant funding hurdles to overcome, if all goes according to plan, the turbines could be up and running before the end of 2007, according to Kristen Burke, wind-siting and community-planning manager for the CWC at the Massachusetts Technology Collaborative. "We know that these projects are very difficult from an economic standpoint, which is why many developers are not looking at them," she says, "but there is over-whelming support for the project in Orleans." Two other towns, Falmouth and Fair Haven (as well as the city of Lynn), are also well into the planning process for projects of their own.[61]

Iowa

The majority of community-owned wind projects in Iowa have been located at public schools. Eight schools currently host ten wind turbines ranging in size from 50 kilowatts to 750 kilowatts. This pattern of develop-ment was the result of a peculiar feature in the state's net-metering law ("net billing" in Iowa), that did not include a maximum cap on the size of the projects. By taking advantage of this provision, the schools were able to install commercial-sized turbines and essentially eliminate their monthly electricity bills. Another factor that has encouraged this model is that many wind-turbine owners have been able to borrow the full cost of their projects at extremely attractive interest rates through Iowa's Alternate Energy Revolving Loan Program (AERLP). The AERLP will provide half of the loan (up to $250,000) at 0 percent interest. The remainder of the financing must come from a private lending institution. These extremely favorable loan provisions, combined with net metering, resulted in some of the loans being paid back in just four to six years. (Not surprisingly, the state's major investor-owned utilities subsequently challenged the open-ended feature of the net billing law, resulting in the establishment of limits, making these types of community projects less attractive.)[62]

Another interesting community wind initiative is the Iowa Distributed Wind Generation Project (IDWGP). A consortium of seven municipal utilities, the project has built a wind farm near Algona containing three Zond Z-750 wind turbines mounted on 50-meter (165-foot) towers. The three turbines combined are capable of producing a total of 2.25

megawatts. The IDWGP was the first utility wind-power consortium in the United States. The $2.8 million wind farm, which began producing electricity in October 1998, was funded by the seven utilities, the Electric Power Research Institute, and the U.S. Department of Energy. Cedar Falls Utilities is the lead utility and project manager, while Algona Municipal Utilities manage operations and maintenance at the farm.[63]

But a description of community wind in Iowa would not be complete without mentioning Waverly Light & Power, a municipal electric utility owned by the city of Waverly, a community of about 9,000 people in northeastern Iowa. In 1993, after receiving grants from the American Public Power Association, WL&P became the first public power system in the Midwest to own and operate wind generation when it installed Skeets 1, a rebuilt used Vestras turbine placed on a farm north of Waverly. In 1999 two 750-kilowatt Zond Z-50 wind turbines, Skeets 2 and 3, were placed on a farm near Alta, Iowa (part of the Storm Lake Wind Facility, which contains 259 wind turbines). Late in 2001, Skeets 1 was retired and replaced with a 900-kilowatt turbine. The new NEG Micon NM52 turbine came online on December 18, and produced over 111,000 kWh from that date to January 1, 2002. The production during those fifteen days was greater than the previous turbine, Skeets 1, produced in an average year. The utility's wind generation serves the equivalent of 761 homes annually.[64] Wind generation contributes 5.5 percent to WL&P's total generation portfolio every year. In May of 2002, NEG Micon USA said that Waverly's wind turbine was among the best producers for NEG Micon turbines in the country. "It was the highest producer of all NEG Micon turbines in the Midwest in May," says Steve Butler, Technical Services Advisor for Waverly Light & Power.[65]

Colorado
In Colorado there is a municipal-utility-based wind initiative located southeast of the city of Lamar. The wind farm, constructed in early 2004, is a collaborative project of Lamar Light and Power and the Arkansas River Power Authority (ARPA). The site consists of four 1.5-megawatt wind turbines manufactured by G.E. Wind Energy. Lamar Light and Power owns three of the turbines and the fourth is owned by ARPA. The three Lamar turbines are estimated to have an annual net generation of 13,809,000 kWh, representing about 14 percent of the city's annual energy needs.

Lamar's wind turbine project was funded by the sale of $6 million of revenue bonds by the Lamar Utilities Board. Each turbine delivered to the site cost a little over $1.3 million, not including construction and site preparation charges.[66]

Michigan

In June 1996, Traverse City Light and Power of Traverse City, Michigan, placed its Vestas V44 600-killowatt wind turbine into operation. TCL&P was the first municipal electric utility in Michigan to install a utility-scale wind turbine. At the time of its installation, the turbine was the only commercial-sized wind generator in the state and the largest in the country, according to the utility. Earlier, some of the residents of Traverse City had decided that they wanted to support renewable energy and agreed to pay a little more on their electric bills, a "green rate" (about $7.58 each month), for electricity generated by this wind turbine located several miles outside of town. The turbine produces about 800,000 kWh of electricity a year, which meets the needs of the 125 residential and business customers who are paying the utility's green rate.[67]

In July of 2000, the village of Mackinaw City, Michigan, began researching the feasibility of constructing wind turbines. The village had excellent wind resources with strong winds coming off Lake Huron and the Straits of Mackinac. The community also had a great location, old sewer spray fields near the wastewater treatment plant, which were no longer used and not well suited for other development. The wind turbines seemed to be a perfect fit. The village worked out a lease and power-purchase agreement with Bay Windpower later in the year, which provided municipally owned buildings with power at a set rate and the village with income from a lease arrangement for the land. The two 950-kilowatt wind turbines went online on December 3, 2001. By the fall of 2003, the turbines had produced over 4 million kWh of electricity.[68]

Wisconsin

Community-owned wind is beginning to gain some traction in Wisconsin, although the state is not blessed with the same wind resources enjoyed by states like Minnesota and Iowa, nor are there as many state policies or incentives in place supporting wind. Nevertheless, a generic "Wisconsin

Community Based Windpower Project Business Plan" was developed in 2003 that offers a variation on the ownership "flip" structures used in Minnesota. Although they are not directly based on this plan, two private but locally owned projects have received all necessary permits for construction. The first is Eden Renewable Energy, LLC, which hopes to install two Vestas NM 82 1.6-megawatt turbines in the town of Eden. The second project belongs to Addison Wind Energy LLC in the town of Random Lake, which plans to install a single 1.6-megawatt turbine on part of a site proposed for a much larger wind project by FPL Energy that was abandoned in 2001 because of strong local opposition. Both of these projects will use a form of the "flip" arrangement, where outside, tax-motivated investors own most of the project for the first ten years to take advantage of the federal PTC, after which the ownership reverts to the local owner(s).[69]

The Northwest

Not everyone in the community-wind movement is following the commercial-scale turbine strategy. In the Northwest region of the United States local wind activists have developed an innovative pilot wind initiative called Our Wind Co-op. A collaboration between nonprofit organizations and utilities, this unique cooperative invests in small-scale wind turbines for farms, ranches, and public and private facilities across the region. Through this collaborative effort, 10-kilowatt turbines are being installed at numerous rural sites serviced by publicly owned utilities in Washington and Montana. The co-op's primary goals are to provide a cooperative model for energy independence, rural economic development, and community ownership. In addition, the co-op is exploring the regulatory, financial, and technical needs of a dispersed but intertied, small wind-turbine network while linking rural producers with urban consumers through cooperative green tag sales.[70] The co-op's first turbine was installed near Peshastin, Washington, in May 2003. Since then nine other turbines have been installed in various locations in Washington and Montana. Our Wind was initially supported by grants from the U.S. Department of Energy's National Renewable Energy Laboratory and the U.S. Department of Agriculture's Rural Development program.[71]

"The public reaction has been overwhelmingly positive," says Jennifer Grove, program director at Northwest Sustainable Energy for Economic

Development, one of the main organizations involved in the project. "For the ten turbines we have put up, I think we had over 350 applications. We get questions from people across the country all the time who would like to replicate our model in their communities, so the program has been quite successful in that respect."[72]

CANADA

Of all the examples of locally based wind projects in North America that follow a cooperative model, the best is Toronto's WindShare. Developed by the Toronto Renewable Energy Co-operative (TREC), WindShare was founded to provide an opportunity for citizens to generate green power in Ontario's deregulated electricity marketplace. The original WindShare project included two wind turbines on the Toronto waterfront. The first turbine, a Lagerwey 750-kilowatt LW 52, is situated at the west end of Exhibition Place and started generating power early in 2003. This $1.6 million, direct-drive turbine, which stands 94 meters (308 feet) high, was manufactured by the Dutch firm of Lagerwey Windmaster B.V, and was the first of its kind in North America. The turbine generates enough electricity every year to satisfy the needs of about 250 homes. The second turbine was originally planned to be erected in another location in Toronto, but may eventually be sited elsewhere owing to a number of issues. WindShare, in a 50/50 partnership with Toronto Hydro, owns the ExPlace turbine and the power it generates. The co-op and Toronto Hydro share equal responsibility for development, capital costs, operation, maintenance, decommissioning, and all other agreements related to the project. The wind turbine that currently makes up this project is the first utility-scale turbine in a downtown urban environment in North America, according to WindShare.

The Toronto Renewable Energy Co-operative was formed in 1998 by members of the North Toronto Green Community, a neighborhood-based environmental group. Their members wanted to create a vehicle for the development of community-based renewable energy in Toronto. With initial grant support from the Toronto Atmospheric Fund, TREC developed the proposal for a community-owned wind turbine located along Toronto's Lake Ontario waterfront. Interest in the project from the public was strong,

WindShare turbine in Toronto, Ontario. Paul Gipe.

and in 1999, TREC created WindShare, which formed a joint venture with Toronto Hydro Energy Services to develop the wind-turbine project. TREC conducted almost three years of wind testing before the site at Exhibition Place was selected. Once the agreements to lease the land, purchase the turbine, and sell the electricity were in place in early 2002, TREC began marketing shares in WindShare, and reached full subscription only eight months after the grassroots marketing effort was launched.

In addition to profiling wind power as a solution to smog and global climate change, the project was also intended to highlight community-based initiatives for the development of renewable energy at a time when government-led initiatives for this approach had been conspicuously lacking. WindShare currently has 427 members that have invested in the ExPlace turbine. Members are entitled to one vote in the governance of WindShare. Voting rights are not based on the number of shares owned, but rather on the basis of one member equals one vote.[73] The co-op model followed by WindShare has been a major reason for the project's success, according to Stewart Russell, former WindShare vice president. "It's been absolutely key," he says. "It's really brought a lot of people together, and I don't think the project would have had the same support without the co-op. The members are really strong advocates, and some of them can't stop talking about the turbine; I know, because I'm one of them. Without the co-op model we would have been just another commercial entity building something on the lakeshore. But since there are over four hundred community activists who are also local owners, I think we have real community acceptance and support. In the summers we even have barbecues at the base of the tower." Another reason for the success of the project was the partnership with Toronto Hydro. "We are very grateful that they have been our partners," Russell adds. "I don't think we could have done it without them."[74]

The fate of WindShare's second turbine is still somewhat uncertain, according to Russell. "We have a large number of people who have invested in a second turbine, but their money is being held in escrow because we have not been able to put up the turbine," he says. "One of the main problems is that shortly after we installed the first turbine, the Ontario government put a cap on electricity prices. We had very low electricity prices here to begin with, and after the cap there was essentially no market for new generation. Also, we have been waiting for what are called 'standard offer contracts' (SOC) to be implemented similar to the German feed-in tariffs, which should be coming out sometime in 2006. This could potentially transform Toronto's community power from a few scattered groups trying to put up turbines here and there to something, perhaps, more like Germany. But until we know what standard offer contracts are going to look like, it really isn't viable for our members to move forward on the second turbine at the present electricity rates."[75]

Russell and other co-op members weren't disappointed. On March 21, 2006, Premier Dalton McGuinty and Ontario Minister of Energy Donna Cansfield announced a dramatic new policy for renewable power projects up to 10 megawatts. The new standard offer contracts (advanced renewable tariffs) were a historic step toward a sustainable energy future, according to the Ontario Sustainable Energy Association (OSEA), an Ontario community-based renewable energy group. "This is a bold step that puts Ontario at the forefront of renewable energy development in North America," says Melinda Zytaruk, general manager of OSEA. "No other jurisdiction in North America has crafted such a striking policy that allows everyone to participate in affecting a sustainable energy economy in Ontario. Standard offer contracts allow homeowners, landowners, farmers, cooperatives, schools, First Nations, municipalities and others to install renewable energy projects up to 10 MW in size and sell the power to the grid for a fixed price for twenty years."

OSEA was instrumental in putting the standard offer concept on the political agenda in Ontario. They launched a campaign in early 2004 to adapt the European policy to Ontario so OSEA's members could form cooperatives to install wind turbines, solar panels, biogas digesters, and small hydro projects. "This type of broad local ownership structure is what we at OSEA call Community Power," says Deborah Doncaster, executive director of OSEA. "A very important aspect of Community Power is that it is locally owned and developed. Community Power has been shown to bring five times more jobs and investments to a local community than projects owned by outside companies." Many of the details of how Ontario's standard offer program will be implemented are to be determined by the Ontario Power Authority and the Ontario Electricity Board. The Ministry of Energy expects contracts to become available by the fall of 2006.[76] The Canadian

Maritime Province of Prince Edward Island implemented a similar, but somewhat more limited, standard offer contract program in 2005.

Assuming that the final implementation of the new standard offer contracts is favorable, WindShare may ultimately decide to install its second turbine in a collaboration with Countryside Energy Co-operative of Milverton, Ontario. In late 2005, WindShare and TREC agreed to collaborate with Countryside Energy to create a new joint entity, Lakewind Power Co-operative, Inc., that hopes to develop a wind farm in Bervie, Ontario, not far from the eastern shore of Lake Huron. The wind farm may contain up to five 2-megawatt turbines and will be a 50/50 cooperative of co-ops, according to Doug Fyfe, the general manager of both Countryside Energy and Lakewind Power. "Countryside Energy Co-operative was a new organization that didn't have any experience installing wind turbines, and WindShare was a co-op with experience that was looking for a better location for its second turbine," he explains. "After some discussions with Ed Hale, who was one of the early leaders of WindShare, we decided to combine our forces, and so far it's proven to be a very fruitful partnership." An agreement has already been signed with two farmer-landowners in Bervie, and a two-year audited meteorological study for the site is available. Assuming all the necessary pieces of the plan fall into place, the Lakewind Power project in Bervie could be operational by the fall of 2008. Two additional Lakewind Power projects are being considered for Milverton and Goderich.[77]

Despite the temporary setbacks with the second turbine, Stewart Russell sees many positive results from the WindShare initiative. "We put the first turbine as close to the provincial legislative building as possible, and a million and a half people see it every day," he says. "It's a really powerful image for people to see a large wind turbine operating in an urban setting. They can go right up and hear that it isn't noisy and see that it's quite graceful with its spinning blades. A lot of people describe it as 'our turbine' even though they are not investors in the project, and there has been great community acceptance. We faced incredible bureaucratic hurdles with this project that most sane folk would have given up on. But we stuck with it, and we ended up with an important asset for our community. This has been a very challenging but extremely enjoyable project to be a part of."[78]

CHAPTER FOUR

WATER POWER

Water power (or hydropower) has had a very long and distinguished history. The oldest machines for capturing the energy of moving water were waterwheels, which date from the time of ancient Greece and Rome. Used by industry in many other countries before the advent of electricity, waterwheels provided the power for grinding grain, sawing lumber, weaving cloth, cutting marble, and many other commercial activities. This hydromechanical power was transmitted from the turning waterwheel to the saws, millstones, and other devices by gears and sometimes cables, shafts, pulleys, and leather belts.

The first industrial use of hydropower to generate electricity occurred in 1880, when sixteen brush-arc lamps were powered by a water turbine at the Wolverine Chair Factory in Grand Rapids, Michigan. That was followed two years later by the first hydroelectric power plant on the Fox River near Appleton, Wisconsin.[1] With the further development of electricity-generating technology, many mill sites were retrofitted with hydroelectric turbines to provide electric lighting and power the mill machinery. Some of these commercial operations sold excess electricity to the surrounding community. In the 1880s and 1890s, many communities with suitable sites constructed municipal hydroelectric generation stations, which provided electricity to the residents for illumination and other household and community purposes such as street lighting and trolley cars (trams). By 1907, hydropower accounted for about 15 percent of the electricity supply in the United States. The industry continued to expand, along with the size of hydroelectric projects, resulting in the abandonment of some of the smaller, less successful earlier ventures. By 1920, hydropower supplied about 25 percent of the nation's electricity. After the stock market crash in 1929, the U.S. hydro industry went through a major shakeout and reorganization, and the federal government

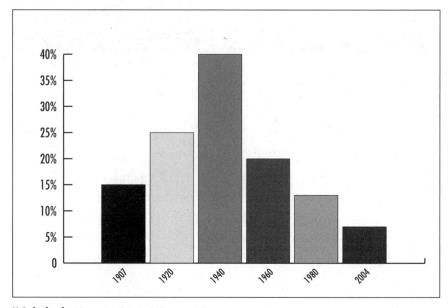

U.S. hydroelectric generation capacity over time. EIA.

emerged as a major promoter of large-scale hydroelectric developments such as the Tennessee Valley Authority and Hoover Dam. After World War II, an increasing number of smaller hydroelectric sites were abandoned in favor of large, centralized generation facilities. In 1940, hydropower accounted for roughly 40 percent of total U.S. electricity production.

Then, during the oil crises of the 1970s, there was growing concern about the nation's dependence on fossil fuels, and a renewed interest in renewable energy—including smaller, local hydroelectric sites. As a result, President Carter signed the National Energy Act of 1978, containing a number of bills, the most significant of which was the Public Utility Regulatory Policies Act (PURPA), which encouraged the generation of electricity from renewable sources. Among other things, PURPA required utilities to pay favorable rates to power producers that used renewable sources of energy. This resulted in a rush to identify and redevelop a large number of neglected hydroelectric sites nationwide. In the state of Vermont, for example, hydropower generation nearly doubled in the years following the enactment of PURPA, with the redevelopment of more than two dozen hydro sites that now meet roughly 6 percent of the state's electrical needs and employ about two hundred people.[2]

WINOOSKI ONE

One of the best examples is the Winooski One Hydroelectric Project, located between the cities of Winooski and Burlington, in northwestern Vermont. This $16.6 million, run-of-the-river project includes a 200-foot long, 35-foot-high concrete dam located immediately downstream of the remains of an old rock-and-timber-crib dam at the falls on the Winooski River. The site had been used for hydropower since just after the Revolutionary War, and the fact that the project was proposed for a site with an existing dam made the permitting process somewhat easier. Unfortunately, this was not enough to ensure quick approval, as the project soon got bogged down in a protracted and bitter struggle between the two cities, local residents, Green Mountain Power Corporation (an investor-owned utility), and the publicly owned municipal Burlington Electric Company.

Enter John Warshow and Matthew Rubin, the two principals of Winooski One Partnership. Warshow and Rubin had been involved in a number of small-scale hydro redevelopment projects in Vermont in the

Winooski One. Winooski One Hydro.

1980s, but Winooski One was probably the most challenging—and, in the end, one of the most successful. In addition to the dam reconstruction, the project also included a concrete powerhouse at the base of the dam containing three turbines with a combined generating capacity of 7.4 megawatts. "There was a huge controversy which had been going on for maybe eight or nine years before we got involved," Warshow recalls. "We were finally able to broker a peace agreement between all of the parties which addressed the many environmental issues and allowed the project to be completed."

In order to accomplish that, Warshow had to secure a series of permits and licenses from the Federal Energy Regulatory Commission (FERC), the state of Vermont, the U.S. Army Corps of Engineers, and the state historic preservation division. In addition, Warshow and his partners spent around $500,000 on a "trap and truck" facility that would capture spawning fish and truck them around the dams, as well as $100,000 to develop a management plan for an endangered plant that grew only at the falls. A riverside park was also proposed as part of the project. Construction finally got underway in December 1991, and Winooski One began generating electricity in April 1993. "We tried to do a little bit for everybody, and the project was not only an engineering success, but a political and environmental success as well," Warshow says. "Virtually everyone has been satisfied with the outcome."[3] Winooski One is a good example of the many problems, and potential solutions, that can result from a collaborative process that involves the community.

HYDROPOWER 101

Hydroelectricity presently accounts for about 20 percent of the world's total electrical supply. Norway produces virtually all of its electricity from its abundant hydro sources, and generates more power per person than any other nation. Iceland comes in second place, with about 83 percent from hydropower. Canada, however, is the world's largest producer of hydroelectricity, covering about 70 percent of its total electrical consumption from its abundant hydropower resources. Despite the fact that the United States is the second largest producer of hydropower, it only generates 6.6 percent of its total electrical needs from hydro as of October 2005.[4] Nevertheless,

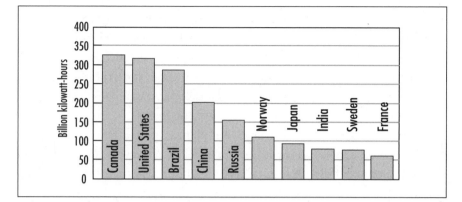

Top hydropower countries. EIA, *Annual Energy Review 1999*, Table 11.15.

the era of large new hydropower dams is essentially over in the United States (and Western Europe), according to many observers. Other countries, however—most notably China, India, and Malaysia—are still actively pursuing large new hydropower projects. China's huge new Three Gorges Dam with a reservoir that extends for almost four hundred miles is one of the most recent—and controversial. Nevertheless, significant potential remains for additional hydroelectric generation in the United States and many other parts of the world, especially for small-scale and micro-hydro installations.

Maximizing hydro development is important, since hydropower plays a crucial role in recovering from a grid failure like the one that occurred on August 14, 2003, in the Eastern United States and Ontario. Hydropower's so-called "black start" capability (the ability to restart generation without an outside source of power) provided the operators of thermal-generation facilities the power to restart their coal and nuclear plants, a process that can take many hours or even days. Systems that had hydroelectric generation available were able to restore service more quickly than those that relied solely on thermal generation.[5] Communities with hydropower resources are clearly going to be in a better position than those that do not as we move into an era of ever tighter energy supplies and an increasingly unreliable national electrical grid. Hydropower can be used to meet either constant baseload power or peak load requirements, and this flexibility is a major advantage.

There are three main types of hydropower facilities: *impoundment,* which uses a dam to store water; *diversion*, which generally uses only a part of a river's flow; and *pumped storage*, in which water is pumped from a lower reservoir to a higher reservoir during periods of low electrical demand and allowed to flow back down when demand is high. Hydropower generation does not produce any emissions or solid waste, but it can cause other problems, depending on the size and location of the facility.

The most common form of hydropower plant is an impoundment facility. In this design, a dam is used to store river water in a reservoir. Water released from the reservoir flows through a pipe or tunnel and then through a turbine attached to a generator, which produces electricity. Although there are about 80,000 dams in the United States, only about 2,400 produce power; the remaining dams are used for a wide variety of other purposes such as flood control or irrigation. Large dams can disrupt fish migration in rivers, displace wildlife and people, and damage habitat. They can also eventually silt up and become unusable. In recent years, large hydropower dam projects have fallen out of favor in the United States, primarily owing to environmental concerns, and it is unlikely that any more will be built. In fact, a movement to remove some dams has gained popularity, which offers some potential environmental benefits, especially for fishing interests. These benefits, however, need to be carefully weighed against the other advantages offered by these facilities. I suspect that the removal of any hydroelectric

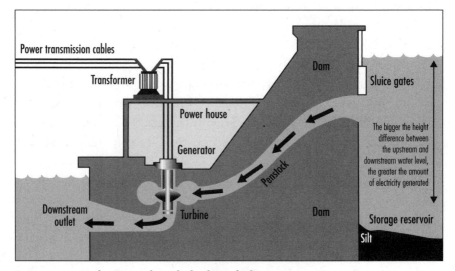

Basic components of an impoundment hydroelectric facility. Environment Canada.

facility will be viewed in hindsight as a terrible mistake as we move deeper into a post-carbon economy.

The diversion strategy (or run-of-the-river facilities, as they are sometimes called) channels a portion of the river through a canal or pressurized pipe (sometimes known as a *penstock*) and does not necessarily require the use of a dam. The water then spins a turbine connected to a generator, and eventually flows back into the river. Diversion hydroelectric installations that don't back up large amounts of water behind them have lower negative impacts than impoundment projects.

Pumped-storage systems are used mainly to even out the load on the grid system, generating electricity when demand is high. Reversible turbine/generators are normally used to pump water one way and generate electricity when run in the opposite direction. Pumped-storage systems have very little negative environmental impact, but do suffer some losses due to evaporation and mechanical inefficiencies.

There are three main sizes of hydropower facilities: large, small, and micro-hydro. Although the parameters can vary somewhat, the U.S. Department of Energy defines large hydropower facilities as those that generate more than 30 megawatts. Small hydropower installations generate between 100 kilowatts and 30 megawatts, while micro-hydro plants produce up to 100 kilowatts.[6]

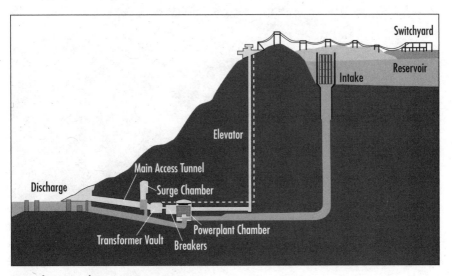

Pumped storage plant. Tennessee Valley Authority.

Aside from the basic requirement for a suitable river or stream, there are two critical factors in any hydropower system, *head* and *flow*. Head is the water pressure created by the vertical distance that the water falls from its intake to the turbine. Head can be expressed as a vertical distance (in feet or meters) or as pressure such as pounds per square inch (psi). The amount or volume of water that is available at the site is called flow, and can be expressed as "volume per second or minute" such as gallons per minute (gpm), cubic feet per second (cfs), or liters per second (lps). Both head and flow must be present to produce waterpower.[7] There are a number of major mechanical components that make up a typical hydropower system. First and foremost is the turbine.

The type of hydropower turbine selected for any site is based on the characteristics of the site. Turbine design is very important and should be carefully matched to the head and flow of the site to maximize efficiency. There are two main categories of turbines, *impulse* and *reaction*. Impulse turbines are better suited for high-head, low-flow applications; reaction turbines are more appropriate for low-head, high-flow sites. Impulse turbines are driven

Pelton-type impulse turbine with housing cover removed. Canyon Hydro.

by one or more high-velocity jets of water and use nozzles to produce the jets. Examples include Pelton and Turgo. Reaction turbines, on the other hand, operate fully immersed in water. Examples include the Francis, Propeller, and Kaplan turbines. A special exception is the *cross-flow* turbine. Technically classified as an impulse turbine because it is not entirely immersed in water, it is, however, used in low-head, high-flow systems. The water passes through a large opening to spin the turbine instead of the small, high-pressure jets in Pelton and Turgo turbines. There are other variations on these main designs.[8]

The drive system, which connects the turbine to the generator, is the next important component. The drive system allows the turbine to spin at its optimum speed, and drives the generator at its best speed for correct voltage and frequency. Ideally this is accomplished by a direct 1:1 ratio connection. But in many situations it is necessary to adjust the transfer ratio to allow the turbine and generator to operate at their optimum, but different, speeds. This can be accomplished by using gears or pulleys and belts. The generator can be either DC (direct current) or AC (alternating current). DC generators are normally only used in small, household systems, usually in conjunction with batteries, and sometimes with an AC inverter. AC generators are used in most larger hydropower systems. A variety of load governors, load management, emergency shutdown, and utility grid-interface controls round out the system.[9]

RECENT HYDRO DEVELOPMENTS

In the United States, especially in New England and the Pacific Northwest, many cities and towns had their own hydroelectric generating plants during the late nineteenth and early twentieth centuries. But with the development of increasingly large, centralized electrical-generation facilities, many of the small community hydroelectric plants were bought up by the large utilities, or eventually abandoned in favor of centralized, large-scale hydroelectric, coal, and nuclear power stations, especially during the 1950s and 1960s when the price of coal and oil was very low.

Although these abandoned hydro sites have been ignored for years, many of them could, and probably should, be redeveloped, since low oil prices are almost certainly gone forever. As I mentioned in chapter 1, approximately 5,677 sites in the United States with a capacity of about 30,000 megawatts have been identified by the U.S. Department of Energy.[10] If these sites were to be developed, they could boost hydroelectric production in the United States by about 40 percent. And since these sites are mostly located on smaller rivers and streams, they offer considerable potential for local, small-scale community initiatives. "I think that DOE estimate is probably conservative," says Daniel New, president of Canyon Hydro. "I look around at the number of streams and rivers that are being utilized today, and there are hundreds of others that are not. I think there is a great deal more potential hydropower available in this country, especially for small-scale projects." Canyon Hydro, located in Deming, Washington, has been designing and manufacturing a wide range of hydroelectric systems for commercial and home residential customers for nearly thirty years.

Fred Ayer, executive director of the Low Impact Hydropower Institute (LIHI), agrees that there is quite a lot of potential in the small hydro sector, but cautions that these projects need to be planned and implemented carefully to mitigate potential environmental damage. "Some people get really excited about hydro because the 'fuel' is essentially free," he says. "While that's largely true, it's not that simple. This is because hydro capital costs tend to be high, and a lot of other people—especially in the environmental community—are not as enthusiastic about hydro for a wide range of reasons."[11] These concerns range from loss of habitat, buildup of sedimentation and bacteria, riverbed and bank erosion downstream of the dam, and disruption of local flora and fauna—especially fish. This has led to extremely contentious battles over various hydropower projects around the country, and in some cases has split the environmental community as well, since hydropower can be sustainable and does not pollute the air. Licensing or relicensing of hydropower facilities have been known to drag on for years—even decades—while the various competing groups fight it out in the licensing process, the courts, and in the media.

LIHI, based in Portland, Maine, is a nonprofit organization dedicated to reducing the impacts of hydropower generation through the certification of environmentally responsible, "low-impact" hydropower. In order to be certified, projects must meet LIHI's eight environmentally rigorous low-impact criteria addressing river flows, water quality, fish passage and protection, watershed health, endangered species protection, cultural resources, recreation use and access, and whether or not the dam itself has been recommended for removal.[12] To date, twenty-one projects have been certified in fifteen states.

Contrary to what many people might think, the size of a hydro project is not necessarily the determining factor in its positive or negative environmental impact. "I've seen very large projects that were designed to minimize impacts, and I've seen very small projects that were complete disasters," Ayer says. A prime example of the latter was located in Columbia Falls, Maine. Ayer was involved in the project's removal. "The Pleasant River Hydro Project was built in the late 1970s," he recalls. "The Pleasant River was one of the few sites where native Atlantic salmon were coming back, but this project blocked the fish migration due to a faulty fish ladder. Unfortunately, it couldn't generate electricity very well either, and the owner went bankrupt. At the time, I worked for Bangor Hydroelectric, and we had an opportunity at another site on a different project where we were causing an impact, and we negotiated with the state and federal governments for mitigation on our project by acquiring the Pleasant River Project and removing it. Although the project had only been in place for five years or so, the damage it had already done was incredible."[13] Nevertheless, Ayer is enthusiastic about hydropower's prospects, especially for smaller, less environmentally intrusive projects.

Another reason for optimism about hydropower in the United States results from a new trend in the licensing and relicensing process—collaboration between some of the previously antagonistic groups. "For many years, there were incredibly unpleasant battles to license and build projects," Ayer says. "Somewhere in the midst of those battles, a desire to find a better approach developed, and people actually began to work in collaborative ways. Increasingly, the resolutions of the difficult issues we struggled with in the past have been achieved collaboratively, and we're finding out that this process enables the community to come up with a satisfactory local solution. We're also finding that the government agencies that are responsible for the regulations and requirements are supportive of collaborative decision making, and they rarely override what the community decision is, as long as it is a truly representative community group."[14]

CLARK FORK HYDRO PROJECT

The Clark Fork Project includes two separate hydroelectric installations on the Clark Fork River of Montana and Idaho owned by Avista, a privately owned energy company (formerly known as Washington Water Power Company). The Noxon Rapids Project, completed in 1958, has a maximum generating capacity of 554 megawatts from its five turbines. The project creates a 7,940-acre reservoir that is 35 miles long and extends upstream to a point near the town of Thompson Falls, Montana. The Cabinet Gorge Project, completed in 1952, has a maximum generating capacity of 236 megawatts from its four turbines. The Cabinet Gorge dam creates a 3,200-acre reservoir that is twenty miles long. This brings the total generating capacity of the Clark Fork Project to 790 megawatts—a large project by almost anyone's standards.

The Clark Fork Project marked the first time in U.S. history that a utility negotiated and reached agreement with a major hydro project's many stakeholders prior to filing a relicensing application with the Federal Energy Regulatory Commission (FERC). This strategy represented a dramatic shift in thinking on the part of the utility and the more than fifty stakeholder groups involved, including five Indian tribes, as well as federal, state, and community organizations. It also resulted in an unusual trademarked process called a "living license," in which the terms of the license were left open for changes as needed. The trust and cooperation generated among the many participants by this strategy was a key element in the successful outcome of the process.

The stakeholder group, composed of one hundred individuals from thirty-nine organizations, formed what became known as the Clark Fork Relicensing Team. The team divided itself into five issue-oriented groups that created what are known as protection, mitigation, and enhancement measures for their respective areas. These areas included fisheries, water resources, wildlife, land use and recreation, and cultural resources. One of the major issues of concern was the protection of native trout. Instead of getting bogged down in endless debate about whether the trout were endangered, Avista agreed to commit substantial amounts of money to restoration of trout watersheds as soon as the team reached agreement.

This was the first time that enhancements were implemented before a hydro project license was issued.[15]

When the collaborative process began, Avista planned on a five-year time line. However, the project was completed two years ahead of schedule; the license application for the Noxon Rapids and Cabinet Gorge dams was filed February 17, 1999, with FERC and approved in a matter of weeks, rather than years. Admittedly, the collaborative process was expensive, since the final agreement required Avista to spend $220 million over forty-five years to protect fish and wildlife, increase recreational opportunities, and address a variety of environmental concerns. Nevertheless, company officials decided that in the long run it would be less expensive to follow the collaborative path, rather than getting involved in years of protracted, and potentially even more expensive, wrangling and litigation.

The collaborative process was so successful that Avista won a Hydro Achievement Award for Stewarding Water Resources from the National Hydropower Association in 1999 and four additional awards in the following years. The company was so pleased with the outcome of the Clark Fork collaboration, that it decided to follow a similar strategy in the relicensing process for its Spokane River hydroelectric facilities.

Noxon Rapids Project. Avista Utilities.

SMALL HYDRO

Large dams have definitely fallen out of favor in the United States in recent years; almost two hundred have been demolished since 1999. While only a few of these dams were ever used to generate hydroelectricity, their removal reflects a general trend toward river restoration initiatives in many communities. Despite the decline in the fortunes of large-scale projects, there has been a recent surge of interest in small hydro nationwide. Driven by high energy costs, a number of federal incentives, and an eased licensing process contained in the Energy Policy Act of 2005, at least 104 small hydro projects in twenty-nine states have been given "preliminary permits" by FERC recently. These projects have a combined generating capacity of 2,400 megawatts, but even more projects are reportedly in the works according to industry observers. "I have noticed a trend of what I call 'incremental hydro,' where people are adding a new generator or a new powerhouse to an existing facility," says Fred Ayer from the Low Impact Hydropower Institute. "There is a great deal of interest and activity in that right now."[16] Most of these recently proposed projects are small—producing less than 20 megawatts of power. But if all 104 projects were built, they would make a substantial contribution to the nation's hydroelectric generating capacity. And if most of the approximately 77,000 dams that do not currently generate electricity were retrofitted to do so, they could generate as much as 17,000 megawatts, according to a recent U.S. Department of Energy report.[17]

The growing interest in smaller hydro projects has resulted in some interesting developments. As I mentioned in the last chapter, municipal (public power) systems in the United States have led the industry in renewable energy production, including wind, solar, geothermal, landfill gas, biomass, and—especially—hydropower. Public power, in fact, leads all other power producers in the percentage of hydropower in its portfolio with over 21 percent, compared to just over 6 percent for investor-owned utilities, and 2.5 percent for co-ops and nonutility generators. This strong emphasis on hydro largely reflects the development of local hydropower sites by communities in the early days of the electric industry.[18] However, since many of these communities developed their hydro sites over one hundred years ago, the potential for additional large-scale municipal projects is rather limited in most parts of the country.

There is, however, one category of community-owned hydropower that has not been widely exploited: municipal water systems. These systems offer a lot of smaller-scale opportunities, according to Johnny Weiss, executive director of Solar Energy International in Carbondale, Colorado. "Often communities have existing infrastructure that could be used or adapted," he says. "For example, Carbondale has a municipal water line that comes down from its source in Nettle Creek. This water line has a lot of pressure that could be used for electrical generation. I think people should look at existing infrastructures and explore what opportunities they might have to make more productive use of them. When they upgrade that infrastructure, that's a good time to think about micro-hydro."[19] In addition to its many course offerings in solar energy technologies, SEI also offers regular workshops on micro-hydro design and installation.

Boulder, Colorado

The city of Boulder, Colorado, has taken advantage of this combination strategy, and eight hydroelectric generators incorporated into the city's municipal water system now generate enough power to provide 11 percent of the electricity needed by the city's 96,000 residents. Five of the turbine-generator units, with a combined rated capacity of 4.1 megawatts, were installed between 1985 and 1987, with one on a raw-water (natural water prior to treatment) transmission pipeline and the other four within the city's water-distribution network.[20] In 1999 and 2004, two more turbine-generators were installed on raw-water transmission pipelines that together provide an additional 6.2 megawatts of generating capacity. A third turbine-generator was purchased from a local utility in 2001 and has 10 megawatts of generation capacity. Boulder's main water supply originates from Boulder Creek in the mountains high above the city and is fed by gravity to the distribution system. Because of the high pressure resulting from the long drop, pressure reducers are needed in the supply lines. The turbine-generators were installed to bypass the pressure-reducing valves. The older generators are rated from 68 up to 800 kilowatts, while the newer units are rated at up to 3.1 megawatts.

From 1990 to 1998, the city's hydroelectric stations generated an average of 15.5 million kWh per year. Including the newer generators, that figure has risen to 42.5 million kWh in 2005. The total installed cost of the first

Municipal water system hydroelectric turbine in Boulder, Colorado. City of Boulder.

five generating plants was $5.2 million, while the cost of installing the additional two turbine-generators was $6.6 million. In 2005, Boulder received $1.7 million from the local electric utility for the hydroelectricity the city produced. Boulder's total revenue from hydropower sales since the beginning of the project is $16.6 million. The environmental impact of the project was minimal, since the water-supply infrastructure was already in place. "That's what makes all of these turbines economical; they are being added to an existing water system," says Carol Ellinghouse, Boulder's water resources coordinator. "And because it's such a huge drop from where we gather the water down into the city, we have to reduce the pressure somehow, and installing a turbine makes sense both from an engineering and financial standpoint."[21]

The city did have one problem with its hydro project, however—finding the turbines and generating equipment. "We've had to look all over the world to get the equipment for these systems," Ellinghouse notes. "We have turbines from England and China and mechanical equipment from Spain. Europe in particular has really moved ahead of the United States with small-hydro technology, and it's really hard to find equipment for this kind of application here in this country." Nevertheless, the hydro project has been so successful that the city is continuing to explore additional hydro potential, and has identified several more sites on municipal water-supply facilities. "It's been a fantastic program," Ellinghouse says of the hydro project. "We've not only made money that offsets the cost of the water supply for our citizens, but we've also offset an awful lot of coal burning that otherwise would have had to take place." Boulder's hydropower project is estimated to have displaced the need to burn 170,000 tons of coal since the first generator went online. Several other cities such as Denver and Colorado Springs have similar municipal water/hydroelectric systems.[22]

The potential in other cities and towns across the nation and around the world is substantial, and a company called Rentricity was formed in 2003 to promote the idea. Headquartered in New York City, the company uses a proprietary system to transform the energy of flowing water (and potentially other materials) in pipes into electricity. A single system can produce between 20 and 300 kilowatts. So far, the company has three projects in Connecticut and Rhode Island and has begun to expand into Pennsylvania.

Washington Electric Cooperative

There are undoubtedly thousands of examples of small, cooperative hydroelectric projects around the world. One, however, owned by a rural electric co-op in Vermont, is noteworthy because it marked the beginning of a move toward a generation portfolio that is somewhat unique. The Washington Electric Cooperative is located in a commercial building in downtown East Montpelier, Vermont. The rest of the mercantile district includes a general store, a post office, a garage, and a video shop. The local volunteer fire department building sits directly across the street from the co-op's headquarters. Despite these modest surroundings, WEC (as it is called by its members) is anything but sleepy. Thanks to a progressive and vigorous management team, in recent years WEC has made a bold move toward greater reliance on renewable sources of electricity at the same time that other utilities in the region were primarily focused on deregulation, restructuring, and short-term power contracts. In early 2001, the co-op's board decided to try to meet its future energy needs through renewables to the greatest extent possible.

In December 2001, the co-op signed a 2.25-megawatt contract with a Connecticut-based landfill to purchase power generated from methane gas.

This enabled WEC to fulfill a long-term goal of divesting itself of nuclear power in 2002 and increasing its renewable sources of electricity to 40 percent of its wholesale power portfolio. The co-op has even higher aspirations, though. This remarkable accomplishment is perhaps unequaled by any other electric utility in the nation. But that's not all. Just a few months earlier, the co-op had received a $1 million federal grant earmarked for wind generation. And in 2003 the co-op's board announced plans to build its own landfill methane-gas-to-electricity project in nearby Coventry, Vermont (more on this in chapter 5). In addition, WEC also buys power from small-scale hydro and woodchip-burning plants and generates 1 megawatt of its own electricity at a flood-control dam in Middlesex.[23] Located on the north branch of the Winooski River, roughly in the center of the co-op's territory that serves its nearly ten thousand customer-members, the Wrightsville Dam hydro project represents the start of WEC's long journey into renewables.

But it was a somewhat rocky start, according to Avram Patt, the co-op's general manager. "The Wrightsville Dam was originally built as a flood control project in 1935 by the Civilian Conservation Corps," Patt says. "The co-op installed the hydropower generating plant in 1984. However, this was done at a time when the co-op went through a fairly lengthy internal political battle between members who were interested in renewable energy and the so-called 'old guard' who were not very enthusiastic about these ideas. The co-op experienced what I call 'the vigorous exercise of the democratic process,' resulting in the approval of the hydro project. But the reason the project happened was not because the board of management at the time necessarily wanted to do it, but rather because they felt the growing political pressure from the members to do it. They responded, somewhat reluctantly, but they did it."[24]

The Wrightsville hydroelectric generating plant is located about two hundred feet downstream of the dam, and a wooden penstock runs between the dam and the three turbines in the powerhouse. The plant is capable of generating about 1 megawatt of power and supplies roughly 5 percent of WEC's total power needs. The plant cost approximately $3.5 million to build. The supporters of renewable energy eventually took over control of the co-op, and also inherited the hydropower facility from the "old guard." The Wrightsville Reservoir created by the dam is now a popular local recreation area, complete with a beach and park, and the co-op contributes to the operating budget of the park as part of the original hydro-plant-approval agreement. Although the Wrightsville hydropower facility is somewhat low-key, it nevertheless plays a small but useful role in the co-op's renewable energy strategy. "It's just sitting there generating electricity as part of our portfolio, and we are happy with it," Patt says.

MICRO-HYDRO

Of the three hydropower sectors, micro-hydro probably offers the most opportunities of all for the development (or redevelopment) of local, small-scale power supply. With an electric generating capacity of 100 kilowatts or less, micro-hydro is well suited for use by homeowners, small businesses, farmers or ranchers, and small communities looking for a renewable supply of electricity. These small hydro systems are similar to their larger relatives in most respects, but are scaled down in size. A 10-kilowatt micro-hydro system can easily provide enough electricity for a large home or small farm, and even smaller systems can provide power for energy-efficient homes or for site-specific tasks like operating machinery, processing agricultural products, lighting, and so on. In some cases a tiny 300- to 500-watt system is sufficient to power most of the appliances in a home (excluding electric heating loads), and with the use of energy-efficient appliances an even smaller generator is possible. Micro-hydro offers the same advantages of larger hydro combined with highly distributed generation, while minimizing or eliminating the disadvantages associated with large-scale, dam-based projects. Micro-hydro systems are relatively inexpensive, can be planned and installed relatively quickly, and are suitable for ownership by individuals, cooperatives, or communities; in short, they are excellent candidates for community-supported energy initiatives.

There are undoubtedly thousands of potential micro-hydro sites that have been overlooked by most state and national planners because they don't offer the kind of size and scale required for the return on investment that most large utilities expect for their shareholders. However, these sites offer a lot of potential for individuals and groups to harness the energy of falling water that is presently being wasted, especially if their main goal is not making a profit, but rather creating a steady supply of electricity to power their basic needs. If the national grid begins to falter or goes down, any local source of electricity is going to start to look pretty good, regardless of its cost per kilowatt hour or return on investment.

Since they tend to be small, most micro-hydro systems use the diversion (or run-of-river) strategy, which does not require a large dam or water storage capacity. This makes micro-hydro even more attractive, since dams tend to be rather expensive engineering projects, beyond the means of most individuals. A typical micro-hydro system diverts river (or stream) water through an intake into some form of water conveyance, usually a canal, which is connected to a forebay that resembles a small rectangular swimming pool. A pressurized pipeline (or penstock) runs downhill from the forebay to a building or shed (the powerhouse), which contains a turbine and other electrical-generation and control equipment. The water from the penstock turns the turbine, spinning a shaft connected to an alternator or generator to produce electricity. After the water passes through the turbine, it flows back into

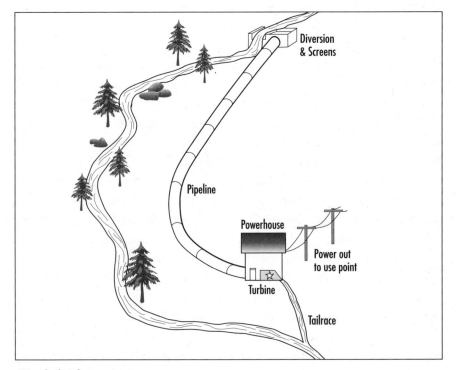

Micro-hydro plan. Canyon Hydro.

the river or stream. The water has not been polluted or otherwise degraded; it's just been temporarily borrowed. In very small systems, some of these components, such as the forebay and the powerhouse, can be extremely basic, or even eliminated altogether. A micro-hydro installation can be a stand-alone system (off-grid) or it can be grid-connected.[25]

Stand-alone systems typically use batteries to store the DC current generated by the turbine-generator. The batteries can supply power at higher levels than the generator capacity and also at times when the turbine is not running during brief maintenance shutdowns. Since only the average load needs to be generated with a battery-based system, the size of the pipeline, turbine, generator, and other system components can be smaller and less expensive than those needed for an AC system. Inverters can be used to transform the DC current supplied by the batteries to AC output for normal household appliances. If the distance from the turbine/generator is short, a 12-volt system can be used. Higher voltages (24, 48, or 120) can be used for longer transmission distances. Most battery-based systems use an automotive alternator to generate the electricity.

In an all-AC system, however, there are no batteries. In this case, the generator has to be able meet instantaneous demand (from motors in refrigerators, freezers, washing machines, water pumps, and so on) as well as the peak load of the entire house, and this generally requires a larger system, in the 2- to 3-kilowatt range. In an AC system, an electronic controller monitors the hydro's output and diverts unused power to a so-called shunt load, such as an electric water heater. The controller acts like an automatic dimmer switch that monitors the generator output frequency and diverts power to the shunt load in order to maintain a constant speed or load balance on the generator. AC systems, however, require more water than DC systems and are more expensive. Regardless of the specific design details, micro-hydro systems are almost always more cost-effective than other renewable power strategies. And once a micro-hydro system has been paid for, the electricity is essentially free, except for minimal maintenance costs.[26]

"Micro-hydro supplies power on a steadier basis than wind or photovoltaics, which is why the system can be smaller," says Paul Cunningham of Energy Systems & Design of Sussex, New Brunswick, in Canada. "The battery bank in particular can be smaller with micro-hydro because you don't have to store power for a week the way you sometimes have to do

Micro-hydro plant. Energy Systems & Design.

with wind or solar. As long as you have a usable source of water, there is nothing else that can provide power at this price, so it's a no-brainer." However, because most micro-hydro systems generate direct current, they can easily be used in combination with wind or PV installations to create an all-season hybrid system. Many small sources of water tend to dry up during the long, hot, sunny days of summer at the same time that PV panels are producing at their maximum, making a hydro/PV hybrid system extremely effective. Cunningham has been in the micro-hydro business for over twenty-five years, and he says that sales have been steadily increasing. He sold around 70 machines in 2003, and sales had more than doubled to 143 by 2005.[27]

Nevertheless, Cunningham admits that, even with a good source of water, micro-hydro is not for everyone, especially for people who are already grid-connected. "There is a certain amount of work and mainte-nance involved, and a lot of people who already have grid power really don't understand how much electricity they consume," he says. "They think they are going to save money with a hydro system, and I have to explain that they are using so much power now, that if we were to provide it with a small hydroelectric system, the interest on the money involved would probably exceed their present electric bill." That's when the initial enthusiasm begins to evaporate, according to Cunningham, who says that a micro-hydro system can cost anywhere from $10,000 to $20,000, depending on a wide range of variables.[28] Then there is the related issue of payback. "Payback doesn't seem to be a big issue for people in most aspects of their lives," Cunningham says. "But as soon as renewable energy comes up, the word 'payback' comes out of nowhere, and that's what they want to know about. And if it's not going to pay them back, then they aren't interested." For off-grid customers, however, who are trying to decide between a $50,000 commercial power-line extension (plus a monthly utility bill) or a $10,000 to $20,000 micro-hydro system (with no monthly utility bill), the hydro begins to look pretty good by comparison.

Finally there is the issue of government support—or lack thereof—for micro-hydro. Many people decide to install a wind or PV system for their home or business because of various state or federal tax credits or other incentives. However, this support is virtually nonexistent for micro-hydro for most people, at least in North America. "This just isn't a level playing field," Cunningham says. "As far as I know, there aren't any subsidies for micro-hydro, but there are for wind and solar. We just got missed because we don't have a vocal political lobby; that's the main reason why this is not happening for us." Consequently, the vast majority of Cunningham's customers are those who live off-grid, and more than 50 percent are located overseas.[29] Micro-hydro projects are especially popular in developing nations, and are being actively pursued in over fifty countries around the world. In Pakistan, for example, over forty isolated villages have installed micro-hydro systems to provide power for small businesses and basic public services, while in Nepal over one hundred hydropower systems have been put into operation. In the United States, community micro-hydro installations are rare.

There are, however, a few isolated, small, mostly rural communities in North America that have installed micro-hydro systems to provide some or all of their electricity. Many of these communities tend to be off the grid (and off the radar screen of public awareness as well—some by choice). Earthaven Ecovillage in Black Mountain, North Carolina, is one example, although community members are not shy about publicity. A self-described "aspiring ecovillage settlement," Earthaven was founded in 1994. Nestled on 320 acres in the forested slopes of the southern Appalachians, the community is located about forty minutes southeast of Asheville. Earthaven currently has about 50 full-time residents, and hopes to grow to a village of at least 150 people on fifty-six homesites. The community's permaculture site plan includes residential neighborhoods and compact business sites, as well as areas suitable for orchards, market gardens, and wetlands. Earthaven, which is 100 percent off the grid, is still a work in progress. The physical infrastructure so far includes roads, footpaths, bridges, camp-grounds, ponds, constructed wetlands, the first phase of a water system, off-grid power systems, gardens, a council hall, a kitchen-dining room, many small dwellings, and several homes.[30]

Lying between 2,000 and 2,600 feet in elevation, Earthaven's forested mountain land consists of three converging valleys with abundant streams and springs, floodplains, bottomland, and steeper ridge slopes. With this diverse physical landscape, the possibilities for micro-hydro were obvious and investigated early on by community members. "We installed the hydro in 1998, and we hadn't built a whole lot else at that point, so I think the hydro was one of the first things we did in the development of the land," recalls Shawn Swartz, a resident of Earthaven and member of the community's utilities committee. "We have two main tributaries here on the property, and the Rosy Branch Creek by far has the most head, so that's where we installed the hydro." The original system includes a 500-watt DC turbine-generator, a charge controller, a bank of four batteries, and an inverter for AC power. Everything, except the batteries, is housed in a small shed. In early 2006, a second 500-watt turbine-generator and inverter were added to the system. "When we built the micro-hydro sta-tion, we provided space for the addition of two more turbines, so it was not that difficult to add the second unit," Swartz says.[31]

The central village area, including the Council Hall, Trading Post Cafe, and White Owl Lodge, operates on electricity generated by the micro-hydro system. Once the second turbine installation is completed, the Village Terraces Co-Housing Neighborhood will be utilizing production from the turbine in a PV-hydro hybrid system. Electricity users on the hydropower system are charged an initial hook-up fee and a monthly per-kilowatt-hour usage fee. The Hut Hamlet Kitchen and every individual hut, residence, or business operates on individual or shared PV electric sys-tems. The community's construction crews use generators to run power tools or get their electricity from the batteries located by the micro-hydro

Micro-hydro at Earthaven. David P. Chynoweth.

station. "When the Earthaven Forestry Cooperative was operating, we built a 'power wagon' to bring power to our construction sites," Swartz says. "This was a small trailer that contained a battery bank and an inverter. We would charge it at the hydro during the evenings and bring it to the job site in the mornings. It could fully power a job site for one to three days, depending on the activities." During a series of rainy, snowy, or overcast days, homes and businesses can run low on electricity, so community members try to use it conservatively. Many rely on propane for refrigeration, while many others use DC-powered refrigeration.

In general, the hydro system has operated well—within its design limitations. "We've had some problems with clogging of the water line or a drop in creek level which resulted in a reduction of power," Swartz says. "And there is no way of knowing what the state of the system is at the point of use, although we do have metering at the hydro shed. So what sometimes happens is the system will crash, the inverter will go into low-voltage alarm and shut itself down. Then, someone has to go over to the hydro building and try to figure out what's happening and get the system going again. The most common problem is that people are just using too much electricity. But, for the most part, the system operates fine."[32]

Earthaven is still actively looking for "hardworking, visionary people of all kinds" to join them in creating their ecovillage dream. For more information, visit their Web site at www.earthhaven.org.

OCEAN ENERGY

The world's oceans can be used for two types of energy: thermal energy from the heat of the sun, and mechanical energy from tides and waves. Thermal ocean energy is relatively constant, while tidal and wave energy is intermittent. This means that somewhat different strategies are needed to harness the energy potential from each source. Oceans are enormous solar collectors: they absorb heat from the sun, which creates wind, which in turn creates waves that produce surface ocean currents. Deep ocean currents, on the other hand, are mainly driven by differences in temperature and water density. In addition, there are ocean tides, which are caused by the changes in gravitational forces exerted by the moon and, to a lesser extent, the sun. And since the seas and oceans combined cover more than 70 percent of the earth's surface, all of this moving water and heat represents huge energy potential if it could be successfully harnessed—a tantalizing goal that has been pursued by many people for decades.

There are three main ocean energy technologies: *wave energy, tidal energy,* and *ocean thermal energy conversion* (OTEC). In the late 1970s and early 1980s, there was a good deal of ocean-energy research and development taking place around the world in response to the repeated oil crises of the 1970s. The United States was no exception. In 1980 and 1981, the

U.S. Department of Energy appropriated over $30 million for ocean-energy systems. Unfortunately, after Ronald Reagan became president in 1981, the tax incentives and funding for ocean-energy projects suffered the same fate as all other renewable energy initiatives in the country, and this promising work came to a halt. Consequently, most ocean energy initiatives in the United States today are small and experimental. The same scenario played out in other countries as well, when low energy prices and pressure from the nuclear and fossil fuel industries resulted in the abandonment of ocean-energy projects of all types.

But there is definitely some interesting long-term potential for this renewable, generally nonpolluting energy source, and in the past few years higher energy prices have sparked quite a lot of renewed research as well as the implementation of some encouraging pilot projects. In January 2005, the Electric Power Research Institute, which has many large utilities as members, issued a report on wave energy conversion in the United States, and concluded that the generation of electricity from this resource may be economically feasible in the near future and warranted further investigation. Three months later, EPRI began the second phase of its ocean research with an investigation into tidal technologies and potential sites in the United States and Canada. This renewed interest in ocean energy by major utilities represents a dramatic shift in thinking on their part, and may herald the revival of the ocean energy industry in the United States.[33]

Another sign of the shifting tide was the revival of the Ocean Energy Council (OEC) in December 2004. Originally organized in 1979, the council promoted ocean-energy initiatives and facilitated communication within the international ocean-energy community. Then, in response to the end of government support for ocean energy in 1981, the council disbanded the following year. After the organization's revival in 2004, OEC and other renewable energy groups and supporters successfully lobbied to include ocean-energy language in the Energy Policy Act of 2005. "We all joined forces to get the language for ocean energy into the energy bill, which now specifically mentions it," says Dan White, editor and publisher of *Ocean News & Technology* magazine. "The language is in the bill, but unfortunately there's no money to support it. Right now, all the money is coming from private sources, which simply isn't enough to cover what needs to be spent, especially for research and development. So, the next step is to try to get the funding." That's no easy task in the United States, where over half the states have no ocean coastlines, and consequently don't have much interest in ocean energy projects, according to White. Nevertheless, White is optimistic about the future prospects for ocean energy. "I'm a huge proponent of this technology, and I believe that ocean energy's time has arrived," he says. "Although it's not really a new technology, a lot of people still don't understand it, so we have a lot of educational work to do."[34]

In other countries, especially those with extensive coastlines, ocean energy has received more private and government support in recent years. As a result, the United Kingdom and Australia presently lead the world in ocean-energy research and technology using wave and current devices. Other nations that are actively pursuing ocean-energy initiatives include Portugal, Denmark, Norway, Italy, India, and China.

Wave Energy

There are three main strategies to harness wave energy. The first uses floats or pitching devices that generate electricity from the bobbing or pitching motion of a floating object. Up until recently these generators have been relatively small, and for the most part only able to power a buoy, beacon light, or other modest electrical device. The second strategy uses oscillating water columns in which the rise and fall of water in a cylindrical shaft causes air in the shaft to spin an electrical generator mounted at the top. The third approach uses wave surge or "focusing" devices constructed near the shore, which channel waves into an elevated reservoir. When the water flows back out of the reservoir it generates electricity using standard hydroelectric methods (a variation on this strategy uses air compressed in a tunnel by the wave action, which spins a generator). Wave-focusing devices were developed in various countries, especially in the United Kingdom in the 1970s, but were abandoned after technical problems and funding cuts. The largest challenges have been to construct devices that can survive storm damage and resist saltwater corrosion. During those early trials, a number of prototype devices were destroyed by storms before they could even be completely installed.

One encouraging recent initiative, however, is the Pelamis wave-energy converter. The Pelamis is a semisubmerged, articulated structure composed of cylindrical sections linked by hinged joints, which makes it look a bit

like a huge mechanical sea serpent. The wave-induced motion of the joints
is resisted by hydraulic rams, which pump high-pressure oil through
hydraulic motors. The hydraulic motors drive electrical generators to produce electricity. Power from all the joints is fed down a cable to a junction
on the seabed. Several devices can be connected together and linked to
shore through a single seabed cable.[35]

In May 2005, Ocean Power Delivery Ltd. (OPD), the Scottish company
that developed the Pelamis concept, announced a deal with an electric
company in Portugal to construct the world's first commercial wave farm.
The 2.25-megawatt project will be composed of three of OPD's distinctive
orange, sausage-shaped, Pelamis P-750 machines located five kilometers off
the Portuguese coast, near Póvoa de Varim. The €8 million ($9.5 million)
project is expected to meet the average electricity demand of more than
1,500 Portuguese households. The project is due to be installed in the fall
of 2006. Subject to the successful performance of the first stage, an order
for an additional thirty Pelamis machines (20 megawatts combined output)
is anticipated. OPD was supported in its research and development work by
€9.8 million ($11.6 million) funding from an international consortium of
European venture capital companies, as well as a subsequent £1.5 million
($2.6 million) investment from the Carbon Trust.[36]

Artist's rendering of a Pelamis wave energy farm. Ocean Power Delivery Ltd.

A somewhat similar strategy that uses a wave-focusing design is being pursued by Wave Dragon Wales, Ltd. The company has been working toward commercialization of the device for over three years, and has deployed a 1:4.5-scale prototype in Denmark since 2003, where reliable power production has already been demonstrated. The demonstrator is a floating, slack-moored wave-energy converter with a rated capacity of 4 to 7 megawatts. The Wave Dragon device allows ocean waves to overtop a ramp, which elevates water to a reservoir above sea level. This creates a "head" of water, which is subsequently released through a number of turbines and transformed into electricity. Water is returned to vents in the base of the unit. The only moving parts are the turbines. The company has plans for a precommercial demonstrator to be located four to five miles off Milford Haven on the southwest coast of Wales in 2007.[37] Expect to hear more exciting news about wave energy in the near future.

Tidal Energy

Twice a day, ocean tides rise and fall in response to the gravitational forces of the moon and the sun. The early history of tidal energy dates back at least a thousand years. In the Middle Ages, small early tidal power plants used the inward and outward flowing seawater to turn waterwheels to grind grain and saw lumber. More recently, harnessing tidal energy generally involved building a substantial dam (called a *barrage*) across a tidal inlet and allowing the incoming tide to flow through a sluice, tunnel, or gateway. When the tide reaches its highest level, the sluice is closed, and the water flowing back out generates electricity using traditional hydropower technology. This strategy is known as an *ebb generating system*, since it relies on the ebb (outgoing) tide. Using the incoming tide to generate power (a *flood-tide generating strategy*) is also possible. Sometimes a gateway is not used and the tide flows unimpeded through the barrage and through the turbines. Two-way generation is another, more recent variation. However, tidal energy systems that rely on barrages can have a negative environmental impact because of reduced tidal flow and the buildup of silt.

BARRAGES

The traditional strategy for harnessing tidal energy has been through the use of large-scale tidal power plants. There are, however, only three commercial-

scale tidal power plants in the world. The largest is located on the Brittany
coast of France, in the estuary of La Rance near Saint-Malo. With a capacity
of 240 megawatts, it generates on the incoming and outgoing tide. The
second tidal plant is located at Annapolis Royal in the Canadian province of
Nova Scotia, with an output capacity of 20 megawatts. The smallest commercial tidal plant is located at Kislaya Guba on the White Sea in Russia,
with a 0.5-megawatt capacity.[38] There are around ten additional barrages
located in various countries, but they are not used for commercial power generation. Only about twenty sites in the world have been identified as potential large-scale tidal power locations.

Construction on the La Rance Tidal Barrage began in 1960. The 330-
meter-long (1,083-foot) barrage, containing twenty-four *bulb turbines*
(named for their shape) each rated at 10 megawatts, was completed in
1967. The barrage also contains a lock to permit the passage of small boats.
The original bulb turbines, developed by Électricité de France, allow generation on both daily ebbs of the tide. These *axial-flow turbines* were also
designed to pump water into the basin to make it easier to anticipate generation levels. This type of turbine is popular with many hydropower installations, and has been used in European dams on the Rhine and Rhone
rivers. The La Rance plant generates about 90 percent of Brittany's electricity. In 1997, the plant's owners embarked on a major renovation and
began to install new turbines that can generate when the tide is flowing in
either direction. This major retrofit is scheduled for completion in 2007.[39]

In Canada, Nova Scotia Power operates the tidal generating station at
Annapolis Royal. Completed in 1984, the Annapolis Tidal Generating
Station was sponsored by the provincial and federal governments, and was
designed to explore the potential of harnessing tidal energy from the inter-
nationally famous Bay of Fundy, home of the world's highest tides (54.6
feet). The tidal facility in Annapolis Royal has the distinction of being the
first and only modern tidal plant in North America. Twice a day the turbine
spins in response to the tides, and the electricity generated is fed into the
provincial electric grid. Annapolis uses the largest *straflo* (straight flow) *turbine* in the world to produce more than 30 million kilowatt hours per year—
enough to power four thousand homes.[40]

Nevertheless, despite the technical success of the La Rance and
Annapolis plants, no major tidal power projects on this scale have been

Annapolis tidal generating station. Nova Scotia Power Inc.

constructed elsewhere on account of a variety of environmental concerns, including navigation restrictions, interference with fish migration and the intertidal zone, and potential changes of tidal patterns downstream. These problems have generated substantial opposition from environmental groups and local inhabitants, essentially bringing barrage-based projects to a halt.

Tidal Turbines

There is, however, another interesting recent development in tidal power that does not rely on environmentally questionable barrages: underwater *tidal* (or *marine*) *turbines*. These turbines (sometimes referred to as *hydrokinetic devices*) are similar to wind turbines, except the blades are underwater and spin in response to the tides. But because tides are more predictable and relatively constant, smaller tidal turbines can produce the same power when compared to larger wind turbines. The marine turbines, which have attracted the attention of the hydroelectric industry recently, can be grouped in offshore underwater farms similar to wind farms or placed in tidal rivers or estuaries. These underwater installations are much less expensive to build than tidal barrages and do not suffer from the major environmental problems associated with them. In addition, there are many more potential sites for underwater turbines, since they can be placed wherever favorable tides are present. Finally, because the majority of the turbine structure (in some cases, all of it) is underwater, it would not raise the visual pollution issues that are generally associated with offshore wind farms.

In May 2003, the world's first underwater marine turbine was installed about one kilometer off Foreland Point near Lynmouth, Devon in the United Kingdom. The 300-kilowatt turbine, mounted on a steel pile set into a hole drilled in the seabed, was the culmination of the "Seaflow" project, a £3.5 million ($6.1 million) initiative conducted by a consortium of United Kingdom and German companies. Since underwater maintenance is difficult and expensive, the rotor and drive train can be raised completely above the water's surface and maintenance performed from a surface vessel. The turbine was designed to achieve its maximum power output with a tidal current of 5.5 knots. The project was designed to provide the necessary operational information needed to build larger commercial systems.[41]

"Seaflow" completed two years of operation in May 2005, setting a new world record for a truly offshore installation of this type. The turbine yielded a wealth of information that has been used to design "SeaGen," a 1-megawatt commercial-scale successor.[42] In December 2005, Marine Current Turbines, Ltd., the Bristol-based designer of the SeaGen, received permission to install the turbine in Northern Ireland's Strangford Lough in 2006, where it will be connected to the national electrical grid. The SeaGen has the capacity to produce enough power for about eight hundred homes. "We have shown that it is possible to generate power in a hostile marine environment and to have a negligible effect on marine life," says Professor Peter Fraenkel, MCT's technical director. "Strangford Lough will demonstrate whether SeaGen has the commercial potential whilst safeguarding the marine environment."[43]

But that's not all. In January 2006, MCT announced preliminary plans for a twelve-unit tidal energy farm in the same general location where the Seaflow test turbine had been operating. The 10-megawatt tidal farm, to be known as the Lynmouth SeaGen Array, would be connected to the local electrical grid and have the capacity to supply clean and sustainable energy to around 5,500 homes in the area. Environmental impact studies need to be conducted in 2006 before MCT decides whether to move forward with

MCT Seagen pile-mounted twin-rotor tidal turbine. Marine Current Turbines™ Ltd.

formal applications for project approval.[44] Additional potential locations are also being considered by the company.

Meanwhile, on the other side of the Atlantic Ocean, a similar pilot project is under way in New York City's East River, where Verdant Power LLC of Arlington, Virginia, has proposed a tide-powered turbine farm that could cost $20 million and generate up to 10 megawatts when completed in 2007. Unlike the SeaGen offshore project in Northern Ireland, however, Verdant's proposed 36-kilowatt turbines will only be about eight feet below the surface of the thirty- to forty-foot-deep river. Nevertheless, if the initial trials of six prototypes prove successful, the underwater farm could grow to three hundred turbines or more ranged along a mile-long stretch of the river. The sixteen-foot-diameter turbines are attached to the tops of large steel pipes anchored in the bedrock, and a yaw system allows them to operate bidirectionally, taking advantage of the ebb and flood tides.

After testing a prototype in December 2002 and January 2003, Verdant received additional funding for six test units to be evaluated over an eighteen-month period that began in the early summer of 2005. These units will become the world's first multi-turbine, distributed generation, underwater array to be connected to the grid, according to the company.[45] In July 2005, the Federal Energy Regulatory Commission made an unusual exception when it ruled that Verdant could continue its tests without a hydropower license. However, the company still needs a wide range of federal, state, and local permits before it can proceed with its commercial-scale plans. Nevertheless, Verdant hopes to complete the final phase of the underwater farm in 2007, assuming that all the permits are issued in a timely manner.[46]

Ocean Thermal Energy

A huge amount of energy is stored in the world's oceans, which absorb enough heat on a daily basis from the sun to equal the thermal energy of 205 billion barrels of oil. If less than 0.1 of 1 percent of this stored solar energy could be converted into electric power, it would supply more than twenty times the total amount of electricity consumed in the United States in a day. If you have ever been swimming in the ocean, you probably have noticed that the deeper you go, the colder the water gets. Unlike tidal energy, ocean thermal energy is relatively constant. Power plants can be built to use this difference in temperature to make energy in a process called *ocean thermal energy conversion* (OTEC). This concept had its roots in the late 1880s, when a French physicist, Jacques Arsene d'Arsonval, first proposed tapping the thermal energy of the ocean. But it was left to others to conduct various experiments to convert that thermal energy to electricity in the mid 1900s. OTEC test projects made a good deal of progress in the late 1970s and early 1980s, but were largely abandoned after the end of government support in 1981.

OTEC makes use of the ocean's natural *thermal gradient*—different temperatures at different depths—to drive a power-producing cycle. As long as there is about a 36°F (20°C) difference between the warm surface water and the colder deep water, an OTEC system can produce a significant amount of electricity. Energy extraction from this temperature difference, however, is difficult and expensive, and OTEC systems typically have low thermal efficiencies of only 1 to 3 percent. In order to make up for this low efficiency, huge amounts of water have to be pumped through the system, requiring a lot of energy, piping, and machinery. OTEC plants can be constructed on land near the shore, on offshore platforms anchored on the shelf, or on moored or free-floating facilities in deep water.[47] In addition to electricity, a 10-megawatt OTEC plant could also produce about 3 million gallons of fresh (desalinated) water a day as part of the process, a real bonus in a time of decreasing freshwater supplies in many nations around the globe.[48]

There are three types of OTEC systems: *open cycle*, *closed cycle*, and *hybrid cycle*. The open-cycle system uses warm surface water to generate electricity. The warm seawater is placed in a low-pressure container, where it boils, creating steam that drives a generator. The steam is then condensed back into freshwater (the salt remains in the low-pressure container) when it is exposed to cold temperatures from deep ocean water. The freshwater resulting from the process can be used for a wide variety of purposes. Closed-cycle systems use a fluid with a low-boiling point, such as ammonia, to spin a turbine to generate electricity. Warm surface seawater is pumped through a heat exchanger where the low-boiling-point fluid is vaporized. The expanding vapor turns the generator. Cold, deep seawater is then pumped through a second heat exchanger, condensing the vapor back into a liquid, which is then recycled through the system. Hybrid systems combine both closed-cycle and open-cycle systems elements.[49] OTEC technology offers a lot of promise, especially for tropical island communities that rely on increasingly expensive petroleum for power generation. There are a number of demonstration projects underway in Japan, Hawaii, and India. Other projects are expected in various locations soon.

FUTURE PROSPECTS

While there has been a good deal of research and some encouraging pilot projects, it remains to be seen whether any of these ocean-energy strategies can be exploited economically on a commercial basis without negative environmental impacts. Ocean energy presently costs more than conventional and other renewable energy sources. Nevertheless, some renewable energy experts are optimistic. Given sufficient funding and long-term support, ocean energy could provide up to 20 percent of the United Kingdom's present electricity needs, according to a report published in 2006 by the Carbon Trust about the nation's wave- and tidal-stream energy sector. The report went on to say that in the nearer term, the sector could meet 3 percent of the United Kingdom's electric supply by 2020.[50]

In the United States, there has been a sudden flurry of ocean-energy project proposals in the past few years, especially since 2005, and a number of industry observers expect the pace of these projects to accelerate considerably in the next few years. So far, there are few if any locally owned or community-based ocean-energy initiatives. However, if offshore wind farms can be cooperatively owned, there is no technical reason why the same general strategy could not be used with marine turbines. Opportunity knocks.

CHAPTER FIVE
BIOMASS

At the homecoming cocktail party beside the family swimming pool in the 1967 film *The Graduate*, Benjamin Braddock (played by Dustin Hoffman) received a word of advice from Mr. McGuire (played by Walter Brooke). That advice was "plastics." In the 1960s, the plastics industry was on the verge of a huge explosion of growth, and had young Ben Braddock taken Mr. McGuire's advice, he might have become a very wealthy man. If the film were made today, the word of advice might very well be "biomass." This is because biomass is about to experience similar explosive growth as a feedstock in the bioenergy and bioproducts industries, leading to a lot of excitement in the agricultural and forest products sectors. Yet most people don't know much about biomass or the huge potential that it offers, so a little explanation might be helpful.

BIOMASS 101

As with virtually all renewable energy strategies, biomass is powered by the sun. Solar energy is captured by plants through the process of photosynthesis, and the resulting organic matter that makes up these plants is known as biomass. Consequently, biomass can be described as "stored solar energy." When they think of biomass, most people think of firewood, our oldest fuel source. Firewood used for cooking and home heating was, and still is, an extremely important fuel in many countries. In addition to these domestic purposes, wood-fired boilers powered steamboats and railway locomotives as well as some industries, especially in the United States, during much of the first part of the nineteenth century. Today, biomass in various forms can be used to generate electricity, and to make solid fuels such as wood chips or pellets for heating or electrical generation (or both).

Biomass can also be used to create liquid biofuels for transportation (described separately in chapter 6) and a variety of other products, including chemicals and bioplastics. The Industrial Revolution was marked by the gradual transition from wood to coal and other fossil fuels. This process is about to be reversed rather abruptly in the near future, with a return to biomass energy sources.

Biomass fuels offer a number of advantages. They are renewable, they have virtually no net carbon emissions,[1] and, if sustainably harvested, they can provide a long-term solution to at least part of our current energy needs. But the principal advantage of biomass fuels is their low cost when compared to increasingly expensive fossil fuels. In addition, biomass fuels and products can reduce the need for oil and gas imports; support the growth of agriculture, forestry, and rural economies; and encourage major new domestic industries such as biorefineries that will make a wide variety of fuels, chemicals, and other bio-based products. And when biomass is efficiently combusted, it also dramatically reduces carbon dioxide emissions when compared to petroleum, natural gas, and coal, making it an attractive strategy to help mitigate global warming. This is because, unlike fossil fuels, the CO_2 emitted when biomass is combusted is taken up by the plants or trees that grow the next crop of biomass fuel, creating a closed-loop carbon cycle. In contrast, when fossil fuels are combusted, carbon that has been locked away for millions of years in the earth's crust is suddenly released into the atmosphere, causing pollution and leading to global warming. Last, but by no means least, biomass strategies are generally scalable, from the individual home system up to the largest industrial-sized facility.

But are supplies of biomass sufficient to meet the upcoming demand? In a 2003 study conducted by Jeff Dukes at the University of Utah, he calculated the total amount of carbon that would be needed to sustain present global energy demand. Dukes discovered that it would take 22 percent of all the plant matter that grows on Earth throughout the entire year to meet our current total primary annual energy requirements (roughly twice our current agricultural use).[2] This offers some sense of the dramatic scale of biomass harvesting that would be required. Nevertheless, in the state of Maine, which is blessed with substantial biomass resources, plans are being developed to give the state's forest industry a boost, while helping to wean the region from its reliance on fossil fuels. The Fractionation Development

Center, a nonprofit based in Rumford, Maine, is pushing a "forest biore-finery" proposal for the state. The FDC recently completed a biorefinery feasibility study that concluded that, within fifteen years, Maine could pro-duce half of the transportation and heating fuel that the state currently consumes. Ultimately, the plan calls for more than sixty biorefineries statewide that would employ around seven thousand people. In addition to producing fuel, the biorefineries could also manufacture a wide array of chemicals aimed at replacing those presently derived from fossil fuels, according to FDC executive director Scott Christiansen. "We have reached the tipping point for significant industrial investment in the renewable energy sector," he says. "The thing that changed was not the maturation of technology—which is forging ahead at breakneck speed—the thing that changed was the acceptance of high energy costs over the long term."[3]

The U.S. Department of Energy (DOE) and the U.S. Department of Agriculture (USDA) have weighed in with their own study. The joint study was prepared for these agencies by the Oak Ridge National Laboratory and published in April 2005. It says that the Biomass R&D Technical Advisory Committee, a panel established by Congress to guide the future direction of federally funded biomass research and development, envisions a 30 percent replacement of the current U.S. petroleum con-sumption with biofuels by 2030. In order to accomplish this, the committee set a very challenging goal: biomass will supply 5 percent of the nation's electrical power, 20 percent of its transportation fuels, and 25 percent of its chemicals by 2030. The goal is equivalent to 30 percent of current petro-leum consumption, and achieving it will require more than one billion dry tons of biomass feedstock annually—a fivefold increase over current con-sumption. The seventy-eight-page study attempts to demonstrate that the United States has the resources—primarily forest and agricultural lands—to meet these goals with sustainable biomass supplies without compro-mising the production of food. Roughly 75 percent of the biomass would come from "extensively managed" croplands, while the remaining 25 per-cent would come from "extensively managed" forestlands. Admittedly, "extensive management" of these lands raises legitimate questions about possible environmental damage, and the potential for this downside must not be ignored. Virtually any large-scale agribusiness strategy has its

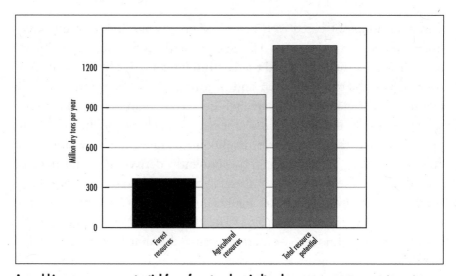

Annual biomass resource potential from forest and agricultural resources. U.S. Department of Energy/U.S. Department of Agriculture.

inherent environmental dangers. In any case, major advances in various technologies, crops, management systems, production yields, and processing—as well as substantial increases in funding—will be necessary to meet these ambitious targets.[4]

While these targets may be optimistic, they are not unreasonable assumptions given a well-funded, comprehensive, coordinated national initiative—which has not been the case so far. But even these ambitious goals emphasize the reality that only relatively modest percentages of current energy consumption can be met with sustainably harvested biomass. Yes, it's possible to meet a portion of that current consumption with biomass, but as a practical matter, it's not possible to cover all of it. This simply confirms what most people in the renewable energy sector have known for years: that it's going to take a wide range of renewable strategies coupled with substantial reductions in demand to meet our future energy needs. It's also important, however, to note that these projections do not take into account the possible negative impacts of accelerated global climate change on crop and forest yields—which could turn these projections on their heads.

In the absence of strong national leadership from the federal government in the United States, an independent, nonpartisan initiative called the

Energy Future Coalition has taken matters into its own hands, and in 2004 launched an ambitious project called 25 x '25. The 25 x '25 Energy Work Group—composed of farmers, ranchers, foresters, and agribusiness leaders—believes that agriculture can help the nation achieve energy independence, and has embarked on a plan to make that happen. The group's vision statement says: "By 2025, America's working lands will provide 25 percent of the total energy consumed in the United States while continuing to produce abundant, safe and affordable food, feed, fiber and fuel." The project has two primary goals: to help these leaders unite behind this vision, and to develop a comprehensive strategy to make it happen.[5] In crafting their statement, the working group felt that agriculture's greatest contribution to the nation's energy solutions would involve the production of liquid transportation fuels (such as ethanol and biodiesel) and the generation of power from wind (particularly from wind farms located on agricultural land), energy crops, and agricultural residues. The group, however, was mindful of the fact that this energy production should take place without negatively affecting food and fiber production. The vision outlined by the working group consequently advocated a food *and* fuel rather than a food *or* fuel production role for U.S. agriculture.

As part of their deliberative process, the group identified three main barriers—an inadequate emphasis on renewables; a lack of strategic vision for agriculture's role in producing energy; and lack of a detailed action plan—and set out to overcome them. The group has also formed a collaborative effort with a number of corporations that expressed interest in the initiative, launched an economic analysis team to review various biomass studies, and began to organize over twenty state-level alliances to build grassroots support. "Renewable energy seems to be a catalyst to bring citizen groups together for a common goal," says Mike Bowman, a fifth-generation rancher and farmer in Colorado, who is coordinating efforts to facilitate creation of the state alliances.[6] To date, well over three hundred agricultural, forestry, and other organizations have endorsed the 25 x '25 initiative.

The 25 x '25 group held its first National Ag Energy Summit in Austin, Texas, in February 2005, where over 140 leaders and energy stakeholders participated in discussions focused on how agriculture could meet its ambitious energy goals. A second gathering in March 2006 took place in Washington, D.C., building on the foundation laid at the previous meeting and providing the group an opportunity to roll out its implementation plan. The momentum generated at the gathering has been growing across the country as the 25 x '25 Energy Work Group has been assisting local leaders to form alliances in more than twenty key states. The group is hoping for endorsements from at least half of the U.S. Congress by the end of 2006. "We've received numerous calls from congressional staff, asking about how they can participate in their districts as well as on the national level," reports cochair Bill Richards. "This issue crosses all party lines and truly captures the American spirit. Whether it's energy from wind, solar, biogas, biomass, or geothermal sources, we can find solutions right here at home."[7]

BIOMASS HEAT

The oldest use of biomass is wood-fired space heating. Because basic wood-heating strategies were developed hundreds (if not thousands) of years ago, they generally are low-tech and have the advantage of working without the assistance of electricity or other modern infrastructure, a real advantage in an energy-constrained future where the national electrical grid may become increasingly unstable. Residential home heating represents the largest share of wood fuel use in the United States. A wide range of wood-stoves, fireplaces, wood furnaces and boilers, as well as masonry heaters can use firewood as fuel to keep your home comfortably warm even in the

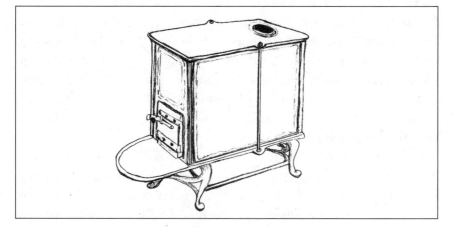

A six-plate cast iron stove from the 1740s. Courtesy of Antiquestoves.com.

coldest of winters. The standard measure of firewood is the *cord* (hence the term "cordwood"). A cord is a measure of volume of a stack of wood that is 4 feet wide, 4 feet high, and 8 feet long.

My personal relationship with biomass heat—and cordwood—has been a long one. Over the years, I have used a variety of wood-fired heating devices, and can personally attest to the fact that they work well, as long as you are willing to expend quite a lot of physical labor on a fairly regular basis, especially if you cut your own firewood. Burning wood has a low environmental impact if it is sustainably harvested and burned responsibly in a well-designed appliance.

Woodstoves

In the United States, all woodstoves sold after July 1, 1992 are required to be certified by the Environmental Protection Agency (EPA). This certification ensures that woodstoves emit less than 7.5 grams of smoke particulates per hour (compared to about 42 grams per hour produced by stoves made prior to 1992). All certified woodstoves now carry a permanent EPA label. The new woodstoves produce almost no visible smoke and deliver efficiencies in the 70 percent range, a substantial improvement over the 40 to 50 percent range of earlier models.

There are many different types and styles of woodstoves that come in a wide range of sizes, from small to very large. Probably the most important factor in choosing a woodstove is determining what size stove meets your home's heating requirements as well as your needs. This is not simply a matter of looking at the manufacturer's claimed heating capacity, because there are a number of other factors involved. In addition to your home's interior layout, you need to consider its level of weatherization as well as the severity of the winters in your location. Also, when it comes to woodstoves, "bigger" is definitely not always "better."

A properly sized woodstove will save you money and provide the best performance. But getting a stove that is too big is worse than getting a stove that is too small. This is because an oversized stove is almost never allowed to burn at its most efficient operating temperature, which can lead to excessive creosote formation and air pollution. An oversized stove will also make its immediate environment uncomfortably hot. At the other extreme, with a stove that is too small, you may be tempted to fire it beyond its capacity, possibly resulting in damage to the stove. A properly installed, high-efficiency woodstove is one of your best hedges against higher fossil fuel prices and uncertain future supplies. Nevertheless, the use of wood-fired heating stoves is restricted in some locations because of air quality issues.[8]

Masonry Heaters

Developed between the seventeenth and eighteenth centuries in northern Europe, the masonry heater is one of the most effective wood-fired heating appliances ever devised, and its outstanding features have withstood the test of time. The masonry heater is also called a *Kachelöfen* or tile stove, as well as a Russian, Siberian, Finnish, or Swedish stove or fireplace. Masonry heaters are still extremely popular in many northern European countries but have not been widely accepted in the United States, due in large part to our over-reliance on fossil fuels. But that's beginning to change.

Masonry heaters are designed for quick, hot fires in their combustion chambers. The heat from these intensely hot fires (1,300°F or higher) passes through a series of baffled chambers (heat exchangers), is absorbed by the large thermal mass of the heater, and then radiates slowly and gently into the surrounding living space for many hours after the fire has burned out. The radiated heat is absorbed by the floor, walls, ceiling, furniture, and even people, creating an extremely comfortable environment with very even temperatures.

Most masonry heaters have doors that are designed to provide optimal air supply for the fire during the burn cycle, which typically lasts about two hours. The rate of air intake is carefully controlled. Except during fueling, the heater doors remain closed, although a few models can operate with the doors open (in which case, the heater looks and functions more like a fireplace). Some doors have glass windows that offer views of the fire while the door is closed. Especially in northern Europe, masonry heater designs are incredibly varied. These heaters range from simple whitewashed clay models, which have served as baking ovens and home heaters, to ornate tiled masterpieces that warmed castles and palaces of European royalty.

Masonry heaters are generally fueled once or twice a day with bundles of relatively small-diameter wood for quick combustion. Heating efficiencies can be up to 90 percent, with very low emissions. The amount of heat produced is controlled by the amount of wood burned. Except when the fire is burning, masonry heaters are silent, so they are a particularly good choice for people who are sensitive to mechanical noise. Masonry heaters work best in homes with open floor plans, but can easily heat individual rooms. Masonry heaters are large, heavy, complicated projects requiring substantial foundations, and should be installed by professional masons with experience in this specific field.[9]

Pellet Heat

In recent years, biomass home heating technology has evolved further, and a large number of wood-pellet and grain-burning heating appliances are now available. Because of their highly efficient combustion process, these appliances have very low emissions and can often be used in areas that otherwise restrict wood-fired appliances. The pellets, generally made out of waste sawdust[10] from the wood-products industry, are normally packaged in forty-pound bags that are fairly easy to handle and store. Because of their homogenous characteristics, pellets are generally used in automated systems. In most locations in the United States you can order bagged pellets by the ton, loaded on pallets. In a few locations, pellets can even be ordered in bulk by the ton, as long as you have a bin in which to store them. A ton of pellets has the heat value of about one and one-half cords of wood (about 20.4 million Btu) and takes up about half the storage space. In some cases, dried corn or other grains can be substituted for the wood pellets, although one can legitimately question the use of potential food as a heating fuel. On account of the dramatic increase in the popularity of pellet-fired appliances in the United States in the past few years, wood-pellet prices have increased as supplies have tightened, but not as much as the prices for fossil fuels.[11]

In northern European countries such as Sweden, Austria, and Denmark, wood-pellet technology is highly advanced and handling systems highly automated. Although other biomass fuels such as wood chips, sawdust, straw, and so on are also utilized, wood pellets are increasingly popular for centralized heating and power generation as well as individual residential heating. In Sweden, wood pellets are often delivered in bulk and are frequently fed into the boiler automatically from a storage bin, and have largely replaced

OUTDOOR BOILERS: BOON OR BOONDOGGLE?

Outdoor boilers (often erroneously called furnaces by their manufacturers) are popular in many rural parts of the United States and Canada for both residential and commercial purposes. Developed in the 1980s, outdoor boilers offer an alternative way to heat your home with wood and other biomass. Outdoor boilers are weatherproof, and some brands even come equipped with an exterior covering that looks like a small storage shed, which protects the boiler's electrical and other components from the elements.

Water is heated in the boiler and then pumped through insulated, underground water lines to your house. The hot water then passes through a heat exchanger or directly into a variety of heat emitters; the water lines can be connected to your domestic hot-water tank as well. A properly sized outdoor boiler can heat not only your home, but also separate garages, sheds, workshops, barns, and swimming pools. Because the combustion takes place outdoors, you won't have to contend with the dirt and dust normally associated with burning wood in the home. However, you will need to don boots, cap, gloves, and parka and trudge out into blowing and drifting snow in the middle of the winter to keep the fire going. But there are other problems.

Because they are located outside, these boilers are exempt from Environmental Protection Agency (EPA) emission standards. Owing to their oversize fireboxes, many outdoor boilers are hard to regulate for a clean burn, and they tend to produce much more air pollution than their indoor cousins. Some manufacturers claim that their units will burn "virtually anything," leading to a lot of irresponsible behavior on the part of owners. There have been numerous ongoing complaints about excess smoke and emissions from outdoor boilers, and their use is banned in some states and many local jurisdictions. In general, outdoor boilers do not measure up to some of their manufacturers' more optimistic performance claims, cause substantial pollution, and are not particularly environmentally friendly.

heating oil and conventional firewood for home-heating purposes. In Denmark wood-pellet boilers provide space heating and hot water for individual houses, and (via district-heating systems) to small communities from one to one hundred families.[12] Denmark has long been a leader in pellet heating appliances, and some of the finest pellet-fired boilers and stoves are manufactured there. With almost half of Austria covered in forests, wood heating is increasingly popular in that nation, and one of the main driving forces has been the development of the wood-pellet-fired boiler for residential heating. Strong government support has played a key role in many of these European initiatives. That kind of support, for the most part, has been absent in the United States.

Pellet stoves, furnaces, and boilers offer the same general environmental advantages of wood-fired heating appliances without a lot of the labor and mess normally associated with firewood. For example, a typical wood-fired stove needs to be tended every few hours, while a pellet-fired stove only needs to be fueled about once a day. A pellet-fired boiler or furnace can be even easier to operate. In the fall of 2005, we replaced the old, oil-fired boiler in our basement with a new Tarm pellet-fired boiler manufactured in Denmark. During the 2005–06 heating season I found that I only needed to fill the pellet bin every two to five days (depending on how cold it was outside). I also had to clean the ashes out of the firebox every three days or so (and out of the boiler tubes every two or three weeks). On average, I probably spent about thirty minutes a week tending the boiler. Anyone who is familiar with wood heat knows that this is not much labor, and I decided it was a fair tradeoff for dramatically reducing our household's dependence on fossil fuels. The fact that we saved over $600 on fuel in the first season by switching from oil to pellets was an added bonus.

The downside of the new pellet-fired heating appliances, however, is that they have augers and blowers that run on electricity. These devices can break down, and in the event of a power failure they stop working altogether, so you need to have some form of backup electrical-generation system or backup heat for an extended power outage (the same is also generally true for standard oil- or gas-fired boilers or furnaces). The strategy in our home is to have a fireplace insert, a woodstove designed to be inserted into a fireplace opening, which burns regular firewood. In an emergency, the insert can heat the entire house.

Tarm pellet-fired boiler in author's basement. Greg Pahl.

But pellet heat is not just restricted to home use. Quite a few businesses have found pellets to be an economical way to heat their buildings and cut their fuel bills at the same time. NRG Systems in Hinesburg, Vermont (which I mentioned in chapter 3), built a stunning new manufacturing and office center in 2004, which included—among its many green features—

two Tarm pellet-fired boilers that are similar to what I have in my own home, only much larger.[13] I visited the 46,000-square-foot, three-story building about six months after its grand opening, and received a personally conducted tour by David Blittersdorf, the company founder. We spent quite a lot of time in the utility room checking out the shiny new blue boilers and then inspecting the thirty-ton pellet storage bin located upstairs. The bin, which holds a year's supply of fuel, allows NRG to receive its pellets in bulk directly from the manufacturer, New England Wood Pellet, located in Jaffrey, New Hampshire, which has experienced exponential growth since the company was founded in 1992. Blittersdorf is a firm believer in peak oil, and he has been working diligently to prepare his company for dramatically higher energy prices as well as shortages of supply.

"We really like the Tarm boilers," Blittersdorf says. "We had to make a few programming changes to the controls to get them to work seamlessly without babysitting, but after that they ran really well and we only had to check them every day or so and clean out the ashes about every two weeks." The pellets cost about half of what propane goes for these days and about 30 percent less than fuel oil, according to Blittersdorf. "The pellets burn very cleanly, and they are easy to store and feed," he says. "We're really happy with the system, and I think pellets should be pushed a lot more than they are."[14] Thanks to its many green features, the NRG building received a LEED gold-level certification, making it the first gold and highest LEED-certified building in Vermont and one of only four manufacturing facilities in the world to receive this designation for environmental and energy-saving design.[15]

Wood-Chip Heat
Burning wood chips is another variation on biomass heating. There has been increasing interest in this strategy in the past twenty years or so, particularly among institutional users, since the equipment for these systems tends to be too large and expensive for most individual homeowners, unless you have a very large house in the country and a number of good-sized outbuildings clustered nearby that also require heat. One limiting factor for these systems is the need for a sustainable source of wood fuel within about a thirty- to fifty-mile radius, since the cost of transportation is a key factor in fuel cost (the actual size of the radius can vary considerably, depending on a wide range of local conditions). The use of wood chips to heat schools is growing in

Michigan, Minnesota, Montana, Vermont, Wisconsin, Arkansas, Georgia, Kentucky, Missouri, Tennessee, Pennsylvania, and Maine. But Vermont is unquestionably a national leader in wood energy, which is not surprising since over three-quarters of the state is forested. One of the most obvious manifestations of that leadership is the fact that more than twenty-five schools in the state have now converted to wood-boiler heating systems. These systems are fired with wood chips or sometimes wood pellets, and are increasingly popular in newly constructed schools or for conversions from high-cost systems like electric heat in existing schools. The savings in heating costs can be dramatic, sometimes by as much as a factor of ten. The Leland and Gray Union High School in Townshend, Vermont, for example, is saving over $40,000 per year with its wood-chip heating system.[16]

But even in Vermont, wood-chip heating does not immediately come to mind when new buildings are built or when old heating systems are being updated, and there has been a concerted effort on the part of several state agencies and the Biomass Energy Resource Center (BERC) in Montpelier (the state capital) to promote this strategy. BERC is the lead partner in the state's Fuels for Schools initiative, in collaboration with three state agencies. A wood-energy expert and other state staff have been working closely with schools to provide them with the resources, information, and advice needed during the new system design and installation process. Many of these schools previously had heated with electric heat, and the conversion process has increased comfort, improved indoor air quality, dramatically lowered costs, and benefited the local economy.

The new Mount Anthony Union Middle School in Bennington had a wood-fired heating system included as part of its construction plans. BERC was the project manager for installing the heating plant, and worked with the architect and engineer to integrate the wood system into the construction project. The wood system began operation in the 2004–05 school year and quickly demonstrated its many advantages. Richard Pembroke, the business manager for the Southwest Vermont Supervisory Union, said the district has been so pleased with the results of the wood-fired boiler at Mount Anthony that it is considering installing one in the high school as well. He said the savings in fuel costs would let the district break even in about a year and a half. "I think that assumption was when oil was at $2.20 per gallon, so I'm not sure it will even be that long," he says.[17]

Schools are not the only Vermont institutions that are taking advantage of wood heat, however. The Addison County Courthouse in Middlebury has a wood-chip-fired system,[18] while many businesses use chips to displace more expensive fossil fuels. What's more, the state office complexes in Montpelier and Waterbury also heat with wood chips in district-heating systems where a central plant supplies heat to a large number of buildings through a system of underground piping (this strategy has been widely used in European countries for many years).

Nevertheless, despite the success of new wood-heat programs in the state, there are already concerns among some forestry experts about the long-term sustainability of these programs. "How renewable is wood in the long run as wood chips become more expensive and less available?" asks David Brynn, director of the Green Forestry Education Initiative for the University of Vermont's Rubenstein School of Environment and Natural Resources. "Vermont's forests produce just under four cords of wood per person in the state every year in terms of growth, but about half of that growth is already being used by the wood industry," he says. "The main point is that we are not awash in wood, even though most of the state is forested. Also, we have to be aware that we will be putting more demands on the forest at the same time that it is being exposed to all the stresses related to climate change: increased storm intensity, stronger winds, drought, and so on. Consequently, as we try to fulfill our increasing needs, we have to be careful not to take what the forest needs for its own health. I think that the best thing that wood can do is to buy us some time while we try to get our renewable energy act together with wind and solar power."[19]

NEDERLAND, COLORADO

In other parts of the nation, wood-chip systems are also becoming more popular. The small city of Nederland, Colorado (population about 1,375), located in the Rocky Mountains about twenty miles west of Boulder, has the first municipal biomass plant in the state. The 30,000-square-foot community center in Nederland is heated by steam from a boiler using wood chips for fuel in this demonstration project that was initiated in May 2003.[20] The secondhand, wood-fired boiler went into service eighteen months later, and replaced the center's two aging natural-gas-fired boilers, which remain as backup for the wood system. The wood chips come from surrounding forests that are being thinned in a fire mitigation project in collaboration with private landowners, the U.S. Forest Service, utility companies, and others. On an average winter day, the facility consumes about three tons of wood chips. Originally, the project was designed to also generate electricity from a microturbine, but the oversized boiler was not able to deliver steam consistently at the required pressure. A future expansion of the community center and increase in heat load might make the microturbine viable. The total cost of the project was $443,000, and the plant is projected to save about $8,150 per year in fuel costs. In addition to saving on fuel costs, the environmental and social benefits of the project include reduction of air emissions (as compared to prescribed burns), use of a renewable fuel source, improvement of forest health, reduction of losses from wildfires, and local economic development.[21]

MOUNT WACHUSETT COMMUNITY COLLEGE

On a larger scale, a new wood-chip heating system is being used at Mount Wachusett Community College in Gardner, Massachusetts. Installed in 2002, the 2.4-megawatt hydronic (hot water) heating plant uses chipped clean-wood waste as fuel. The new heating system, which contains state-of-the-art technology in chip transports, wood-fired boilers, emission controls, and energy management systems, replaced the college's expensive electric heating. The system uses about 1,000 tons of wood chips during the heating season to heat the 427,387 square feet of building space on the college's campus. The wood-fired system has saved about $350,000 in energy costs per year, according to Robert Rizzo, the director of facilities administration. "Overall, the new heating system has worked extremely well," he says. "From October to the end of March it just runs smoothly without having to be shut down." Rizzo admits that initially one high-tech part of the emissions-control system did not work properly, but that once it was replaced with a more traditional bag-filtering system the problem was resolved. "The financial savings of the wood-heating system are substantial, and we see it as a valuable aid in local and regional economic development as well. We think there could be more systems like this installed in other locations."[22] That may very well happen, since the college's bioheat plant has already become a demonstration site and model for other institutions to study.

The wood-chip-fired boiler at Mount Wachusett Community College. Mount Wachusett Community College.

While biomass heat is not for everyone, it does offer a viable alternative to fossil fuels in many instances and consequently deserves serious consideration as long as a sustainable supply of wood fuel is available. The subjects of wood and biomass heat are vast, and entire books have been written about them. My 2003 book, *Natural Home Heating: The Complete Guide to Renewable Energy Options* (Chelsea Green) contains much more detail for homeowners, and I recommend it if you are considering biomass (or other renewable) heat strategies. David Lyle's *The Book of Masonry Stoves: Rediscovering an Old Way of Warming* (Chelsea Green, 1996) is another excellent source on traditional wood home heating.

BIOPOWER

Biomass electrical energy (also called *biopower*) can come from trees, crop residues such as corn stover and wheat stalks, or from crops grown especially for energy production such as switchgrass. Switchgrass, a summer perennial grass that is native to North America, is resistant to many pests and plant diseases, and capable of producing high yields with very low applications of fertilizer, making it an ideal biomass feedstock. Improved varieties of switchgrass have also been developed for livestock forage. Biopower can also come from organic wastes, such as methane gas from landfills and methane digesters on farms or even at municipal sewage treatment plants. Methane is a potent greenhouse gas with about twenty-three times the negative impact as CO_2; landfills produce methane gas from the decomposition of organic matter. A system of buried pipes in the landfill collects the gas, which is then fed through various filtering devices, compressed, and used as fuel in a modified natural-gas engine-generator set, which produces electricity.

Farm methane digesters are devices (usually a closed tank) that use *anaerobic bacteria* (microorganisms that live and reproduce in an environment that does not contain oxygen) that feed on organic matter (usually cow manure) and produce methane gas as a waste product. The methane gas is then piped to a modified natural-gas engine-generator set, which produces electricity. An additional benefit of the anaerobic digestion process is that it kills virtually all pathogens in the manure, significantly reduces the odor of the waste, and improves manure management. Although these systems tend to be more cost-effective on large farms, smaller farms could still make use of methane digesters, especially if their main goal is to produce electricity for their own use, rather than selling it to the local utility. In Europe, farm-based methane digesters are nothing new; farmers in Denmark and other countries have been converting manure into energy for more than fifteen years.

Urban waste streams also offer significant biogas potential. One estimate suggests that if all the municipal wastewater sites in large urban areas of Ontario, Canada, were to use anaerobic digesters and converted the methane to simply generate electricity, this would produce 1.5 GWh per day. This strategy is applicable almost anywhere. A United Nations Development Programme study says that community biogas plants could be one of the most useful decentralized sources of energy supply. Nepal has recently overtaken China and India in the number of biogas plants on a per capita basis. Each of its 125,000 digesters prevents the equivalent of five tons of CO_2 from being emitted into the atmosphere every year.[23]

There are four main types of biopower: *direct-fired; cofiring; gasification;* and *small, modular systems.* Most biopower plants use the direct-fired method, which has systems similar to fossil-fuel-fired power plants. They burn the biomass feedstock directly to produce steam, which spins a turbine connected to a generator that makes electricity. While steam generation technology is proven and dependable, it has limited efficiency, in the low 20 percent range. Biomass boilers are also generally smaller, in the 20- to 50-megawatt range, versus coal-fired plants in the 100- to 1,500-megawatt range. Cofiring is a strategy used mainly by coal-fired electric-generation plants to reduce emissions, especially sulfur dioxide, by mixing coal and bioenergy feedstocks. Cofiring is the most economic near-term option for introducing new biomass power generation, and because much of the existing power plant equipment can be used without major modifications, cofiring is far less expensive than building a new biopower plant.[24]

Gasification of wood and charcoal to produce a low-energy-content synthesis gas (syngas) became popular during World War II as a way to power a variety of vehicles and industrial machinery. More recently, gasification systems are being used to convert biomass into syngas, which can then be burned in a gas turbine or internal combustion engine that spins an electric generator. This offers advantages over directly burning the biomass. The biogas can be cleaned and filtered to remove problem chemical compounds, and can be used in more efficient power generation systems called combined-cycles, which combine gas turbines and steam turbines to produce electricity. The efficiency of these systems can reach 60 percent. Taking advantage of these efficiencies is a major goal of research programs in the United States and the European Union.

Small, modular systems can be fueled by any of these methods and are used in small-scale community, commercial, or individual installations. These systems are now actively under development and can be most useful in remote areas where biomass is abundant and electricity is scarce. Farm-based and community biomass systems, in particular, offer a lot of potential for local, cooperative initiatives. Biomass recently surpassed hydropower as the largest domestic source of renewable energy and currently provides over 3 percent of the total energy consumption in the United States, according to a joint U.S. Department of Energy and U.S. Department of Agriculture study.[25]

McNeil Generating Station

The Joseph C. McNeil Generating Station of the Burlington Electric Department (BED), located in Burlington, Vermont, was the largest wood-fired electric-generating station in the world when it began operating in 1984. Today, with a generating capacity of 50 megawatts, it's still the largest utility-owned, wood-fired plant in the United States. The facility is located about a mile from Lake Champlain in the so-called Intervale, the green floodplain of the Winooski River that flows through Burlington.

Built to replace an aging 1950s-era coal-fired plant, the McNeil Station was designed to make use of a local, renewable, cost-effective, nonpolluting fuel source—wood chips—to simultaneously keep more energy dollars circulating in the local economy, improve the conditions of the state's forests, and provide jobs for local residents. On all these counts, the strategy has generally been successful, according to John Irving, the McNeil plant manager. "It may not have done it in some of those early years to the degree that we expected it would because, back around 1986, the price of fossil fuels plummeted and wood chips were not nearly as competitive," he says. "We were anticipating that the plant would burn 500,000 tons of chips a year, and there were many years where we didn't come close to that amount. But at this point, we estimate we are approaching $200 million that we have put into the state economy since the plant was built."[26]

Final construction costs were $67 million ($13 million under budget), and the plant was even completed ahead of schedule. The McNeil Generating Station is jointly owned by BED (the operator and 50 percent owner), Central Vermont Public Service (20 percent), Vermont Public

Power Supply Authority (19 percent), and Green Mountain Power (11 percent).[27] BED, a publicly owned municipal utility, has long been a national leader in efforts to promote not-for-profit electric rates, local control, and sustainability.

According to BED, 70 percent of the wood chips that fuel the McNeil Station are whole-tree chips that come from low-quality trees and logging residues that are cut and chipped in the forest and then transported by truck directly to the station or to a remote railcar loading site in Swanton, Vermont, located about thirty-five miles north of Burlington.[28] The majority of this chipped wood comes from privately owned woodlots. A BED forester monitors each harvest to make sure it is conducted properly and in accordance with strict environmental standards. The remaining portion of McNeil's wood requirements are met by purchasing residues such as sawdust, chips, and bark from local sawmills, and by using clean urban wood waste (significantly reducing the volume of wood waste going into the regional landfill).[29]

Half of Vermont's forest inventory is wood that has no potential for manufacturing quality products such as woodenware or furniture, according to figures published by the U.S. Forest Service. This commercially unusable wood consists largely of poorly formed trees and treetops left behind after trees have been harvested as sawlogs or pulpwood. The amount of wood available for whole-tree chip harvesting has been conservatively estimated at one million green tons per year in northern Vermont alone. This is twice the amount required to operate the McNeil Station. At full load, approximately seventy-six tons of whole-tree chips (about thirty cords) are consumed by the station per hour. In 1989, the capability to burn natural gas was added to the station, but in recent years, owing to the high cost of this fossil fuel, very little has been used, according to Irving. The station also has the ability to burn fuel oil, or any combination of fuel oil, gas, or wood, but wood remains the primary fuel.[30]

The station has not been without its problems. Because it was sited immediately adjacent to a residential neighborhood, there were complaints about odor, noise, and ash. The station also initially relied on long-term fuel contracts at the insistence of lending agencies. These contracts, however, caused big headaches when the station's capacity factor was reduced in its second year of operation on account of New England Power Pool

Experimental wood gasifier at the McNeil Generating Station in Burlington, Vermont. National Renewable Energy Laboratory.

requirements,[31] causing a huge oversupply of wood chips to accumulate, resulting in odors and spontaneous combustion of the decomposing mountains of chips. Most of these issues have since been resolved, and the station—as well as biopower in general—has a bright future, especially with higher oil and natural gas prices, according to Irving. "It's all driven by fuel prices," he says. "Back when McNeil was built it was because oil was $25 a barrel, and the plant was overwhelmingly voted in by Burlington residents. Everybody was switching to wood heat in their homes as well. Then, in the mid 1980s, when oil got cheap again, people didn't want to mess with wood anymore. But now that the price of oil is going up again, there is renewed interest in wood and other alternatives. It's too bad that we have to wait for high oil and gas prices to make renewable fuels popular, but that certainly seems to do it."[32]

Vermont has a second wood-chip-fired power station located in the small rural community of East Ryegate near the banks of the Connecticut River on the eastern edge of the state. This privately owned facility, built in 1992 at a cost of $46 million, has a generating capacity of 20 megawatts and burns about 250,000 tons of chips annually.

VERMONT GASIFICATION PROJECT

But that's not the end of the story. The McNeil Generating Station has been a magnet for a variety of interesting experiments. One of the most significant was the Vermont Gasification Project (VGP). This large-scale research/demonstration project, constructed at McNeil, converted up to two hundred tons of wood chips per day into a gas (sometimes called syngas) that was used to cofuel the station's wood-fired boiler. The VGP was a collaboration between utility, federal and state government, and private industry participants to refine gasification technologies. Biomass gasification involves converting various biomass fuels to create a relatively clean combustible gas. The advantages of biomass gasification include its ability to convert relatively inexpensive stocks, such as sawdust, switchgrass, agricultural wastes, or energy plantation crops like willow trees, into fuel that will not produce as many emissions, especially of greenhouse gases, as will the direct burning of biomass fuels. These advantages are enhanced when biomass gas is used to fuel an energy-efficient gas turbine.

The process used in the McNeil gasifier was a *circulating fluidized-bed* system, invented by Battelle Laboratories of Columbus, Ohio. In 1980, Battelle began testing the process on a pilot scale, connecting to a 200-kilowatt gas turbine at its Columbus labs. Simply stated, here is how it works. Conveyors transport the biomass from hoppers into the gasifier. At the same time, sand heated to 1,500°F (815°C) is fed into the gasifier along with pressurized steam to form a fluidized bed of rapidly circulating material that distributes heat throughout the biomass. This causes the biomass to break down into a medium heating-value gas and a small amount of char that mixes with the sand. The char, sand, and gas are then separated. The char is sent to a combustor. The heat from the combustor is used to reheat the sand, which circulates in a closed loop.[33]

The process was licensed to the Future Energy Resources Corporation (FERCO) of Norcross, Georgia, in 1992. In 1994, the U.S. Department of Energy and FERCO jointly funded the construction of the biomass gasification demonstration at the McNeil station under the auspices of the DOE's Biomass Power Program. Construction of the gasifier was completed in 1997. After the gasifier was finally started up, the operators made a number of process improvements to various materials-handling and gas-scrubbing systems. The gasifier sometimes supplied between 10 and 15 percent of McNeil's fuel needs. Although the gas was normally mixed with wood chips, it was burned alone at one point during a temporary interruption to the station's chip supply.[34] The project shut down in 2001 owing to lack of funding, and was followed by the bankruptcy of FERCO several years later. Nevertheless, the project did meet some of its original goals, according to McNeil plant manager John Irving. "The gasifier worked," he says. "It took us longer to prove that it worked than we thought it would and it cost us more money, but the main reason the project was halted was financial, not technical."

Many DOE-funded projects like the Vermont Gasification Project tend to rely heavily on federal dollars in the initial stages, and more on private investment in the latter stages. The VGP was approaching the critical funding shift point when the 9/11 attacks took place in New York City, and private funding for speculative projects like the VGP quickly dried up, according to Irving. "The project was never expected to be commercial," he notes. "It was strictly a demonstration project designed to show that the technology worked at this scale, and we actually made good progress."[35]

The VGP was originally designed in three stages. The first was to build the gasifier and show that it could work. "Basically, we did that, and it worked fairly well," Irving says. "We took the gas from the gasifier and burned it in the McNeil boiler, and in the end we got it to run pretty reliably. The next step was to take the gas and clean it up so it would be suitable for use in a gas turbine, and that was what we were in the process of doing when they pulled the plug on the program. We were making some progress, but we weren't there yet, so we never got to the third phase of using the gas in a turbine." The gasifier is currently unused, and its fate is uncertain. The main elements of the gasifier are too large and heavy to transport to a different site, and the cost to dismantle the facility has been estimated between $1.5 and $2 million.[36] I'll have more to say on gasifiers in a moment.

WILLOW CROP EXPERIMENT

But the biomass experiments at McNeil have extended even further. In 1997, Burlington Electric Department joined forces with the Salix Consortium[37] to participate in a multistate biomass plantation project based on research and extensive work in Sweden, the United Kingdom, and Canada on so-called coppice harvest cycles. The idea was to adapt traditional European planting and harvesting systems to North American conditions. As part of the project, BED foresters planted a test plot of more than five thousand hybrid willow trees in Burlington near the generation station. The purpose of the test plot was to enhance research already conducted by the consortium regarding growth, pest resistance, and soil and nutrient requirements of these plantings. The Vermont planting evaluated the performance of various willow varieties in soil and climate conditions different from those studied in New York. "I think there were seventeen different varieties that we tested, and only one of them didn't work out because it turned out to be a favorite food for deer—they ate all of them," John Irving says.

The quick-growing willow trees can be harvested every three years. After the first growing season, when the trees were dormant, BED foresters cut them back to stumps to encourage the sprouting of multiple stems that boost growth, and three years later the willows were harvested. The first harvest was conducted in November 2000, and yielded about thirty-five

A wood-gas powered bus in Germany. Biomass Energy Foundation.

tons per acre. "We took that wood and dried it and then fed it into the gasifier the next spring and it worked okay," Irving says. The trees can grow in
soils that are not otherwise suitable for farming and may eventually prove
to be an alternative crop for idle farmland. The trees can also be planted
along stream banks to act as filters for nutrients and loose soil that cause
water quality problems in rivers and lakes.[38]

Small Modular Systems

Discovered at the dawn of the Industrial Revolution, syngas (short for "synthesis gas," also known as wood gas, producer gas, gengas, biosyngas, and so
on) was originally used to power stationary engines. In addition, cities used
to have gasworks that generated "town gas" (a type of syngas made from
coal) that gave rise to the "gaslight" districts of many cities around the
world. But most people today are unaware of the fact that between 1920
and 1949, wood gas produced from a wide range of biomass fuels was also
used to power cars, trucks, tractors, trolleys, trains, boats, and even motor-
cycles. In addition, large wood-gas electrical generators that ran on wood,
coconut shells, or other waste were used in tropical areas before World War
II. In the Ivory Coast and Gabon a small number of these are reportedly
still in operation.[39] During World War II, in particular, wood gas was used

for many industrial purposes such as sawmills, rock crushers, pumping stations, and even fishing trawlers when normal gasoline and diesel fuel supplies were cut off. Today, syngas can also be used to fuel boilers (like the McNeil Station in Vermont during the Vermont Gasification Project) for steam and electricity generation or as an intermediate feedstock for the production of methanol and other synthetic liquid fuels (more on that in chapter 6). In addition, the hydrogen from syngas can be used to produce ammonia, which is valuable as an agricultural fertilizer. The hydrogen portion of syngas could also be used to power fuel cells. While wood gas is valuable in a wide range of industrial processes, its use in small modular systems probably offers the most interesting potential at the local level, where it can even be used for cooking and heating.

During the oil crises of the 1970s, there was a flurry of interest in syngas/wood gas, and millions of dollars were spent on research projects in the United States and elsewhere. But with the return of cheap oil and cuts in federal support for renewables in the 1980s, most wood gas research ended. However, there was quite a lot of valuable research literature generated at that time, much of it based on earlier work from the 1940s and 1950s. These out-of-print resources have been collected and preserved by Dr. Thomas Reed of Golden, Colorado, who has worked in the field of biomass fuels since 1974. "That was the first distillation of the enormous amount of literature that was available at the time," Reed says. "That's mostly what I kept and am now publishing."[40] Reed's Biomass Energy Foundation Press offers over twenty-five rare titles on biomass energy and biomass gasification. Reed, a physical chemist, is widely viewed as one of the world's preeminent experts on syngas/producer gas.

Reed has also been actively involved in gasification research and development projects for many years, and in 1999, he collaborated with Community Power Corporation of Littleton, Colorado, in the development of a 5-kilowatt "turnkey, tarfree" gasifier. CPC now has about a dozen 15-kilowatt versions of this gasifier in the field and in use by the U.S. Forest Service and others to turn forest litter and biomass trash into heat and power. The company is also developing a broader line of gasifiers for a variety of electricity, heat, and shaft-power applications, and as components for integration with emerging distributed generation technologies such as microturbines, fuel cells, and thermoelectric generators.[41] Reed

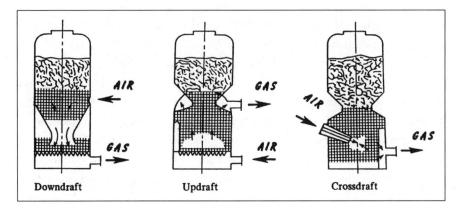

The three basic types of generator. E.E. Donath. From *Producer Gas: Another Fuel for Motor Transportation*, Noel Vietmeyer, ed. (Biomass Energy Foundation).

continues his research and development work on gasifiers at the Hydrogen Science Foundation in Golden, Colorado. The foundation currently has a gasification project in a remote village of about seventy-five households in Brazil. The system uses fruit pits to generate hydrogen-rich gas to produce electricity that powers the fruit processing plant by day and provides electricity for the village at night.[42]

Wood Gas

Wood gas can be generated from a wide range of solid fuels, including wood, charcoal, coal, peat, and agricultural residues (we'll skip coal and peat and focus on the renewables). Here's how it works. Wood gas is made when a thin stream of air passes through a bed of glowing coals from any number of biomass fuels. The fuel can be in its raw form or compressed into bricks or pellets (the same wood pellets used to fuel pellet stoves and boilers work well). The gas is generated in a small gasifier—a metal tank with a firebox, grate, air inlets, and an outlet for the gas produced. On the incandescent carbon surface of the glowing coals, most of the carbon dioxide and steam initially formed by the burning fuel are reduced to carbon monoxide and hydrogen. When mixed with air, these gases are combustible in standard spark-ignition (normally gasoline-powered) engines, and (if clean enough) in gas turbines. Wood gas will also work in diesel engines with the addition of small amounts of diesel fuel.

Best of all, wood gas generators are not rocket science. They are relatively easy to make and can be constructed in small machine shops equipped for sheet metal, welding, and steel-pipe work. Common, everyday materials such as mild (low-carbon, or "soft") steel, standard pipe components, filters, gaskets, springs, and latches are required for the majority of the parts, although a stainless steel "throat" is useful. No major modification of existing spark-ignition engines is required—only a minor replacement or attachment to the carburetion system is needed.[43]

The system that produces wood gas has four main components:

1. A generator to produce gas from the solid fuel
2. A cleaner to filter out soot and ash from the hot gas
3. A cooler to condense tars and other liquid impurities
4. A valve to mix the wood gas with air, as well as a throttle valve to meter the mixture into the engine intake manifold.

The Generator

The generator is the heart of the system. Typically, it is a cylindrical or rectangular metal tank containing space for fuel, a firebox, and an ash chamber. The upper part holds the fuel, normally a thirty-minute to two-hour supply. The lid can be opened for refueling and is often spring-loaded to allow for the relief of any pressure that might build up inside. The fuel falls into the combustion chamber by gravity. Air drawn through the firebox section keeps the fire burning and produces a bed of red-hot charcoal that is at least six inches deep and sufficiently compact that the gas streams flowing through must contact the glowing carbon surfaces.

There are three main types of combustion chambers, which differ in the relative positions of the air inlet and the gas outlet used: *updraft, downdraft,* and *cross-draft.* In updraft generators, air enters below the firebox, passes upward through the glowing charcoal, through the raw fuel above, and exits near the top. The emerging gas has practically no ash in it, but it does contain tars and water vapor picked up as the gas passes through the unburned fuel. Updraft generators are the simplest to build and operate and are best suited for tar-free fuels such as charcoal and for use with stationary engines.

In downdraft generators, air enters the firebox above the fire zone, and the combustion gases then pass down through the hot charcoal and exit near the bottom of the generator. This is a good design for vehicle use and for wood fuels because the impurities are carried into the fire zone, where tars are degraded and steam reacts to produce water gas (a process by which hydrogen is produced by combining steam with coke gas). A constriction in the hearth (the "throat") helps to ensure that all the gaseous products pass through the hottest zone. Downdraft generators produce much less tar, but more ash, in the gas than updraft generators.

In cross-draft generators, the air enters through a nozzle projecting into the side of the firebox. The gases travel horizontally through the hot coals, and exit through the opposite side of the generator. This type of generator is suitable for use with dry, low-tar fuels. The use of wood as a fuel requires the use of either a cross-draft or downdraft generator. Over the years, there have been many variations on these basic designs.

The Filter/Cleaner

Before it enters the engine, wood gas must be filtered to remove the ash and soot. Failure to remove these impurities may result in excessive wear, carbon deposits, sludge in the oil, and possible seizing of the engine. Most early filters were cumbersome, used felt, horsehair, sawdust, cork, or steel wool, and needed to be changed frequently. Today, fiberglass is a simpler substitute. A device known as a cyclone separator, which whirls ash and soot out of the hot gas stream, can also be included in the system. A variety of modern filtering and scrubbing systems can reduce these wood gas impurities to levels below those contained in gasoline.

The Cooler

To remove various other impurities remaining from the combustion process, an air cooler is usually placed after the filters and cleaners. The cooler condenses the liquid contaminants and cools the gas before it is piped to the engine. The cooler the gas, the more power it provides because the amount of combustible material in a given volume is increased. Coolers can be fairly small and simple, especially when used on a vehicle, since the vehicle's motion promotes efficient cooling.

Mixing Valve/Carburetion

At this point, about 70 percent of the total heat value of the wood gas remains available for complete combustion in the engine. But for the wood gas to burn, air must be added. Normally, the pipe that brings the gas from the cleaning and cooling system to the engine is attached directly to the engine intake manifold, bypassing the carburetor. A small valve attached to the pipe allows air to enter and mix with the syngas. A second valve controls the amount of gas-air mixture that enters the engine, and acts as a throttle to control engine speed. The pipe is also often fitted with a simple valve to release any pressure that might build up to prevent flash-backs along the pipe toward the generator.[44]

LIMITATIONS

There are some limitations to this design, however. In most internal combustion engines, wood gas produces less power—as much as 50 percent less in some cases. If used on a vehicle, the added weight of the gas generator further reduces vehicle performance (this is why stationary installations are generally more practical). Wood gas is much less convenient to use than liquid fuels; the fuel is bulky and difficult to store and handle. Maintenance, operational training, and operator experience are required to keep a gasifier running properly. When used on a vehicle, before it can be operated, a fire must be lit in the generator, which takes from two to twenty minutes. If the vehicle stops for a few minutes, the driver has to decide whether to keep the fire lit. On long trips, about twenty minutes are needed to refuel and service the generator every 125 miles (200 kilometers). The ashes must be removed from the generator and the filters changed or cleaned regularly. The tars inside (similar to creosote on the inside of stovepipes and chimneys) are smelly and sticky and contact with skin needs to be avoided owing to the presence of carcinogens. Finally, wood gas itself can be hazardous, since it contains carbon monoxide—a tasteless, odorless, colorless, and highly toxic gas.[45]

INTERESTING POTENTIAL

Clearly, wood gas in this small-scale, low-tech form is an imperfect strategy. But in the absence of affordable gasoline and diesel fuels—especially in an emergency—it does offer some interesting potential, especially for stationary installations in locations with significant biomass resources. Sweden recognized the potential value of wood gas as a backup strategy

Dr. Thomas Reed standing next to a 1948 Chevrolet pickup truck converted to wood power by Mel Strand. Biomass Energy Foundation.

many years ago, and is already prepared for such an emergency. The nation has built and tested gasifiers of three standard sizes: plans are available, machine tools and stampings are designed, and technicians have been trained to manufacture them.[46] A great deal of gasifier development work is taking place in other nations around the world. Gasifiers are currently manufactured and in use in a number of countries such as Singapore, China, Russia, and South Africa. However, India probably has the largest number of small gasifiers in operation of any nation, according to Tom Reed. "They have been doing a lot more with this than we have, although at a lower level," he says. "They generally don't ask as much of the technology, so it's much cheaper."[47]

Even in the United States, a small but dedicated number of wood-gas enthusiasts have built generators and installed them on vehicles. In 1979, a Chevrolet Malibu station wagon was driven 2,700 miles from Jacksonville, Florida, to Los Angeles, California, fueled entirely by scrap wood. In 1981, a firewood-powered pickup truck sponsored by *Mother Earth News* magazine was constructed in North Carolina and subsequently driven around the country by enthusiastic magazine staff members. More recently, Mel Strand of Boulder, Colorado, built a gasifier-powered truck. During World War II, Strand found himself trapped in Nazi-occupied Norway, where he drove a gasifier truck—delivering groceries by day, and weapons to the Underground by night. After his return to the United States, Strand had a long career in machining and fabrication. As a hobby project, he decided to try to recreate the gasifier he remembered from Norway, and installed it on a beautifully restored 1948 Chevrolet pickup truck. A few years ago, Tom Reed went to see the truck. "Mel turned on the auxiliary starting fan, and started the gasifier on large aspen chunks with a newspaper," Reed says. "After a few minutes he lit the gas at the front of the truck and started the engine. We drove around Colorado Springs for several hours. I realized what an art it was to drive a gasifier car, since he could control the spark advance, air/fuel ratio and throttle, all from the steering column, while talking about the old and new days."[48]

BIOGAS

But there are even more ways to produce biopower. The Environmental Protection Agency (EPA) requires all large landfills to install collection systems to prevent landfill gas—mostly methane—from building up and causing an explosion, and to minimize its release into the atmosphere. The productive use of this landfill gas (LFG) to produce electricity is an increasingly popular strategy that makes a lot of sense, although it's not really renewable, since the gas in any given landfill will eventually run out. One can also wonder whether large-scale landfills will survive in an energy-constrained future where virtually all resources will be reused or recycled out of necessity. Nevertheless, LFG still represents a useful biopower fuel source at the present time. According to the U.S. Department of Energy, there are more than 350 landfills nationwide that tap into this source of energy. Transforming what would otherwise be a waste product that contributes to global warming into green energy is also a way for utilities to shift their energy portfolios away from nuclear and fossil fuel generation.

The Coventry Landfill
Vermont's Washington Electric Cooperative, located in East Montpelier, unquestionably had these issues in mind when it constructed a 4-megawatt power plant that burns methane collected at a landfill in Coventry,

Caterpiller engine-generator set at the Coventry gas-to-power plant. Washington Electric Cooperative.

Vermont, not far from the Canadian border. The $8.4 million project, which also included the construction of 7.2 miles of new transmission line to connect with the statewide transmission system, began generating electricity in July 2005. The 7,000-square-foot powerhouse at the landfill presently contains three big, yellow Caterpiller engines that turn the generators to produce about one-third of the co-op's total power requirements. And once a planned landfill expansion takes place, the co-op will install a fourth engine by 2008, boosting total power production to 6 megawatts, or 50 percent of the co-op's projected load.[49]

The Coventry Project was the culmination of three years of planning and hard work by WEC's staff, board of directors, consultants, lending agencies, state regulators, and landfill operator Casella Waste Management, as well as seven months of construction by contractors and linemen in and around the landfill site. In June 2004, after the project had received state approval, the co-op's members blessed it with a special vote of 1,633 in favor to 86 against. The WEC project was the first significant new electricity generation project of any kind to open in Vermont since 1997, when Green Mountain Power launched its 6-megawatt wind farm in Searsburg. And the timing couldn't have been better. The long, hot summer of 2005—coupled with rising oil and natural gas prices—caused the wholesale electricity market to sky-rocket. Prices for electricity on the spot market (daily purchase) averaged 11 to 17 cents per kilowatt-hour (kWh), and actually reached 45 cents/kWh for a few hours in August. "We started production just at the height of the summer," says WEC President Barry Bernstein, "when energy costs were going up with the heat. The plant came online in time to lessen our exposure to those prices." The Coventry power plant now provides the co-op with electricity at a cost of less than 5 cents per kWh during the initial twenty-year contract period.[50]

Foster Brothers Farm

Another example of biopower is the methane digester on the Foster Brothers Farm in Middlebury, Vermont, a fifth-generation dairy farm with about 1,900 owned and rented acres and 370 milking cows. In the 1970s, concerns about manure management and energy supply motivated Robert Foster and other members of his extended family to look into the possibility of installing a manure digester on the farm. "My motivation was to take what most people viewed as a negative and turn it into a positive opportunity," Foster says. "I read about and visited Mason Dixon Farms' digester in Pennsylvania, and over a period of years we received a number of proposals for a digester for our farm. One would have required seven acres of tanks. Needless to say, we didn't use that plan."[51]

The Fosters eventually ended up working with a construction firm that had received a Department of Energy grant for a digester pilot project on a forty-five-cow dairy farm in New Hampshire. "We refined a digester design, got our permits, and built a system in 1982 that fit our needs," Foster says. The digester is an eight-foot-deep, underground concrete tank with a flexible cover on top, which allows it to expand or contract in size as necessary. Anaerobic bacteria decompose the manure in the tank, and this results in the release of methane. The methane is then transferred by a pipe to another building, where it is scrubbed to remove hydrogen sulfide gas and then fed into an internal combustion engine. The engine is connected to a generator that produces the electricity used by the farm. The cow-powered system generates about 360,000 kWh of electricity per year, and the original idea was to sell it to the local utility. But the price the farm received for its electricity eventually declined to the point where it was no longer economic, and the Fosters decided to use their electricity to power the farm and three adjoining residences.

In 1992, the Fosters took their idea one important step further. In addition to producing enough electricity to power the farm, they decided to compost the solids remaining from the digestion process. This resulted in the production of soils, composts, and growing mixes, and the Fosters created a sister company, Vermont Natural Agricultural Products (VNAP) to market and sell its composted soil products throughout the Northeast. VNAP grossed $1.75 million in sales in 2005, while the dairy side of Foster Brothers grossed about $1.2 million in milk sales.

In 1999, after obtaining federal money from the Department of Energy, Foster Brothers Farm reconfigured its system to become a research facility to study the use of methane digesters for the Vermont Methane Project. The project, which ran from 1999 to 2004, was designed to explore more effective new uses of methane recovery technology on Vermont dairy farms. During this time, one successful experiment used the system to create steam to transfer heat to the digester, increasing its efficiency and demonstrating that a smaller tank size was possible.

"Some of this technology's advantages, like odor control and greenhouse gas reductions, are a little hard to put numbers on, but I think people should consider taking a look at it. In any case, we definitely achieved our original goals," Foster says.[52]

Cow Power

The Vermont Methane Project also resulted in the construction of a larger farm methane digester at Blue Spruce Farm in Bridport, Vermont, which has about 1,500 head of cattle. The farm, owned by the Audet family, received a $97,318 USDA Rural Development Renewable Energy grant to purchase and install an electric generator for its methane digester—a large, 14-foot-deep, concrete bunker that measures 72 by 100 feet. Blue Spruce, which now generates 1.75 million kWh a year, has become the flagship project for Central Vermont Public Service Corporation's popular Cow Power program. This voluntary program allows CVPS customers to sign up for farm-methane-generated electricity by paying an extra 4 cents per kilowatt-hour on their monthly electric bills. Customers can choose to buy 25 percent, 50 percent, or all of their electricity through the Cow Power rider. Customers using 500 kWh per month who choose to receive 25 percent of their power under the rider would only pay an additional $5 a month.[53]

This option has been popular with many CVPS customers, and over three thousand five hundred have signed up, making Cow Power one of the most successful programs of its type in the nation.

Around a dozen other Vermont farms have been seriously considering the idea, and in April 2006, four more decided to join the program. Farms in Sheldon, Fairlee, West Pawlet, and St. Albans are receiving grants totaling $666,000 from CVPS's Renewable Development Fund, to defray the cost of building new farm-based electric-generating systems to support the state's largest renewable energy program. "These grants will help develop 8,400 megawatt-hours of clean renewable energy right here in Vermont," CVPS President Bob Young said. "That's enough energy to supply 1,395 average homes using 500 kWh per month." The farms all hope to be online by the end of the year. [54]

"We believe Cow Power has something most renewable programs don't have: a close link between the customer, the energy, and the local benefits," Young says. "Lots of companies are offering renewable programs, but often the energy is generated several states away, and there's no real relationship between the energy and the local environment. With Cow Power, many of our customers know the local farm-producer, and there is a sense of partnering to better the environment. That's what makes Cow Power special, and it's one of the things we believe will help it grow quickly."[55]

COMBINED HEAT AND POWER

Combined heat and power systems (CHP), also known as *cogeneration*, generate electricity and thermal energy in a single, integrated system. This contrasts with the typical model in the United States, where electricity is generated at a large, central, coal-or gas-fired or nuclear power plant, while heating and cooling needs are met by small, on-site systems in individual buildings—a highly inefficient and wasteful strategy. Because CHP captures the heat that would otherwise be lost in traditional separate generation of electricity, the efficiency of these combined systems is much greater—more than twice that of plants that only generate electricity. And when a CHP system is also renewably fueled, the advantages multiply. CHP has become an extremely popular strategy in European countries such as

Denmark, Finland, the Netherlands, the Czech Republic, and others.[56] Much of this CHP is coal-, oil-, or natural-gas-fired, but a growing number of CHP projects are biomass-fired.

Biomass CHP projects are particularly effective when they are part of a district-heating system. District-heating systems use one central plant to heat a large number of residential or business structures within a reasonable distance of the plant with hot water or steam via insulated, underground heating pipes. A district-heating plant can provide higher efficiencies and better pollution control than a larger number of smaller boilers.

Denmark, for example, has fifty wood-chip-fired, twenty-five wood-pellet-fired, and seventy-five straw-fired district-heating plants. Twenty of these are CHP installations. Although many of these are pilot projects, the Danish government is extremely optimistic about the potential for expansion of the biomass CHP sector.[57] In 2000, Finland had forty-eight locations in the country with CHP production connected to district-heating systems. The capital, Helsinki, has the largest CHP system, with 1,017 megawatts in electric capacity and 1,300 megawatts in heating capacity. CHP plants produce about 76 percent of Finnish district-heating energy. While these plants use a variety of fuels, including oil, coal, natural gas, and biofuels, a recent EU study concluded that there was potential for an additional fifty-one plants that could be fueled entirely with biofuels.[58]

It is ironic that, in the late 1890s, CHP systems were the most common generators of electricity in the United States. As the cost and reliability of the electric power industry improved, users abandoned their on-site electric generation in favor of more convenient purchased electricity. During the oil crises of the 1970s there was renewed interest in CHP, but many utilities refused to purchase excess power from CHP facilities, limiting on-site electricity generation to what could be used there.[59] Renewable portfolio standards, net metering, and other regulatory advances in recent years have changed all that, and CHP strategies have begun to catch on again.

One of the most dramatic examples of combined heat and power is in St. Paul, Minnesota, where the city is using wood waste to both heat and cool most of its downtown buildings while also generating electricity at a facility owned and operated by St. Paul Cogeneration. The CHP plant, which went online in May 2003, burns about 280,000 tons of chipped wood waste every year, and feeds 25 megawatts of electricity into the Minnesota power

St. Paul, Minnesota, CHP plant. Market Street Energy.

grid. The heat from the plant meets about 80 percent of the annual energy needs for District Energy St. Paul, Inc., a company that provides district heating and cooling services to the majority of buildings in the city's downtown. The CHP plant is jointly owned by Market Street Energy Company, a District Energy St. Paul affiliate, and Cinergy Solutions of Cincinnati, Ohio. Cinergy Solutions designed and built the plant, which includes a unique combination of renewable energy, CHP, and district-heating technologies. St. Paul Cogeneration is the largest wood-fired combined heat and power plant supplying a district energy system in the nation.[60]

The plant has completed its initial break-in period, and although it experienced the usual startup issues for any large new facility, it has already achieved many of the goals that its designers had hoped for. "We have been able to achieve quite good performance with the plant, and we now have a system that is really considered to be a model for others to follow," says Mike Burns, vice president of operations for Market Street Energy. "This is the final phase of what we had envisioned for a community energy system using a renewable energy source in a very efficient combined heat and power process, and it's enabled us to maintain very low and stable energy rates here in downtown St. Paul. Our experience so far has been that the wood waste fuel has cost us about a quarter of what natural gas would have cost, so it's been quite economical to burn wood." In addition, there has been strong community support for the project from both community leaders and grassroots organizations in the area, according to Burns.[61]

Through the use of wood-waste fuel and the CHP process, District Energy has reduced its reliance on coal by 80 percent, soot emissions by 50 percent, and greenhouse gas emissions by more than 280,000 tons annually. The use of wood chips as fuel for the CHP plant also provides a greater amount of sustainability to the community and helps solve a wood-waste disposal problem.

There is one major limitation to the whole concept, however, according to Burns. "There is only a limited amount of biomass available, and if the idea becomes too popular and you end up with too many biomass-fired plants in any given area it can create a problem with the supply of bio-mass," he explains. "It's a great idea as long as the biomass-fired plants are situated far enough apart so they aren't competing for the same limited bio-mass feedstocks." The plant typically draws its fuel from a seventy-mile radius around St. Paul, but occasionally obtains some of the biomass from as far as several hundred miles away.[62] This problem of limited fuel supply again emphasizes the reality that biomass is only a part of the renewable energy solution picture, and that conservation will have to be a major part of that picture.

A BIOGAS-POWERED TRAIN

In a rather unusual biomass variation, the world's first biogas-powered train, named "Amanda," carries passengers on the 80-kilometer (50-mile) run between the Swedish cities of Linköping and Västervik. The Tjustbanan rail line links Linköping in south-central Sweden with Västervik on the country's east coast. While the biogas-powered train is noteworthy in its own right, what's really unusual about this project is that the biogas comes from the entrails of dead cows. While the train's mileage isn't great—only about 2.5 miles per cow—the concept is intriguing. The methane-powered train, operated by Swedish Railways, gets its fuel from a local biogas plant that turns cows (or at least parts of cows) into fuel. The cows are slaughtered for meat anyway at a local abattoir, so the use of the

Biogas powered train in Sweden. Lasse Hejdenberg, Svensk Biogas.

entrails for biofuel is a unique way of turning one business's waste into another's feedstock—a concept that's central to sustainability. This strategy also solves the problem of how to dispose of the offal, which can no longer be processed into animal feed owing to concerns over mad cow disease. The cow parts are hauled from the slaughterhouse to the biofuel plant, Svensk Biogas, where they are essentially stewed gently for a month before the methane is drawn off. While the methane fuel for the train does cost about 20 percent more than conventional diesel fuel, the price differential is expected to shrink as the prices for oil and natural gas continue to rise.[63]

The train was developed by Svensk Biogas, which converted an old 1981 Fiat rail diesel car by replacing its diesel engines with two Volvo gas engines. The train is equipped with eleven gas cylinders containing enough biogas to run for 600 kilometers (375 miles) before needing a refill, and it can reach a maximum speed of 130 kilometers (81 miles) per hour. The train, which cost Svensk Biogas 10 million kronor ($1.3 million) to develop, went into service in September 2005.[64] But the biogas initiative in Linköping is not restricted to the train; the city's sixty-five buses are also powered by the fuel. In fact, the city claims that it was the first in the world to run its bus fleet on methane. Taxis, garbage trucks, and a number of private cars also fill up at the biogas pump.[65] Sweden, in fact, is a world leader in gas-powered vehicles, and has around 4,500 natural gas vehicles, 40 percent of which run on biogas that is produced in community biogas plants.[66]

Obviously, if cow guts were to become the fuel of choice for the world's transport system, there wouldn't be enough cows—and milk drinkers would not be very happy either. But Sweden is determined to have the world's first oil-free economy, and the biogas-powered train and municipal vehicles in Linköping and elsewhere are some of the many strategies the nation is pursuing to achieve this ambitious goal by 2020. Other nations would do well to follow Sweden's example.

CHAPTER SIX

LIQUID BIOFUELS

Solar, wind, hydro, and geothermal energy are all important renewable strategies; however, their usefulness in transportation is, at best, limited. While a wind-powered sailing ship is nothing new, it's hard to envision a practical wind-powered dump truck, or a water-powered school bus. Approximately 185 billion gallons of petroleum was used for transportation fuels in the United States in 2004. Since transportation is probably going to be affected the most by peak oil, liquid fuels made from biomass will play the most significant role in this sector for the immediate future (electric vehicles may play a supporting role in local transport). This is why liquid biofuels are so important—and why they are attracting so much attention.

Although the term "biofuels" can refer to a wide range of biomass-based fuels, the term generally refers more specifically to *liquid* biofuels. The primary liquid biofuels are ethanol (sometimes referred to as bioethanol), biodiesel, methanol, and to a lesser extent, butanol and propanol. Ethanol and biodiesel are the most widely used and currently receiving the most attention, so we'll focus mainly on them. Each of these fuels can be made in different ways from different feedstocks. Each has its advantages and disadvantages as well as its limitations. Before we get into the details, a little background might be helpful to put all of this into context.

BIOFUELS HISTORY

Although biodiesel is a relatively recent development in the United States, the basic chemistry involved has its roots in the nineteenth century. As early as the mid-1800s, *transesterification* (the process now used to make biodiesel) was used as a strategy for making soap. Early feedstocks were corn oil, peanut oil, hemp oil, and tallow. The alkyl esters (what is now called biodiesel) resulting from the process were originally considered just a nuisance by-product. Ethanol also had a place in early U.S. biofuels history mostly as lamp oil, but it was not until after 1906 (when an alcohol tax was finally repealed by Congress) that demand for ethanol as a fuel began to grow, especially for internal combustion engines.[1]

A number of Europeans are credited with inventing the internal combustion engine, especially Nikolaus August Otto, who produced an early version around 1860 in which he used alcohol as a fuel. Rudolf Diesel, the German inventor of the engine that now bears his name, was also a firm believer in the potential of biofuels for powering his engine. Although his early prototypes burned a wide variety of fuels, an engine built on his design that was entered in the 1900 Paris Exposition ran on peanut oil to demonstrate its fuel flexibility. But as relatively cheap petroleum-based fuels became more widely used in the early twentieth century, Diesel's engines were modified to burn a low-grade petroleum-based fuel, which later became known as "diesel fuel." Nevertheless, Diesel remained a strong advocate for the use of vegetable oils as engine fuel. In a 1912 speech, about a year before his death, Diesel said, "The use of vegetable oils for engine fuels may seem insignificant today, but such oils may become, in the course of time, as important as petroleum and the coal-tar products of the present time."[2]

In 1896 Henry Ford built his first automobile, a quadricycle, to run on ethanol. In 1908 Ford's famous Model T was designed to run on ethanol, gasoline, or a combination of the two. Ford was so convinced that the success of his automobile was linked to the acceptance of "the fuel of the future" (as he described ethanol) that he built an ethanol production plant in the Midwest. He then entered into a partnership with Standard Oil Company to distribute and sell the corn-based fuel at its service stations. Most of the ethanol was blended with gasoline. Ford's biofuel turned out to be fairly popular, especially with farmers, and in the 1920s ethanol represented about 25 percent of Standard Oil's fuel sales in that part of the country. In retrospect, Ford's alliance with Standard Oil may not have been such a good idea. As Standard Oil tightened its grip on the industry, it focused its attention on exploiting its petroleum markets—

and eliminating any competition. Nevertheless, Ford continued to promote ethanol through the 1930s. But finally, in 1940, he was forced to close the ethanol plant because of stiff competition from lower-priced petroleum-based fuels.[3]

During World War II, because of the disruptions to normal oil supplies, virtually all the participating nations made use of biofuels to power some of their war machinery. At the same time, transesterification used in the soap-making process became the subject of great interest because of the by-product *glycerin* (or glycerol), which is a key ingredient in the manufacture of explosives. But after the war, the return of steady oil supplies and low gasoline prices brought an end to biofuels production in the United States, and for all practical purposes the industry ceased to exist. It was not until the oil shocks of the 1970s that biofuels began to experience a renaissance.

The price increases and fuel shortages of the 1970s and early 1980s spurred interest in the development of alternative fuels around the world, but especially in Western Europe and the United States, where the economic damage from the turmoil in the petroleum markets was most severe. In 1979, in response to the ongoing international oil crisis, ethanol-gasoline blends were reintroduced to the U.S. market, and promoted as "gasoline extenders" and octane enhancers. The modern ethanol industry evolved out of these initiatives in the United States and abroad.

While ethanol was being promoted as a gasoline additive, a separate but related line of research into alternative diesel fuels made from vegetable oils was being conducted in a number of countries. Two long-term research programs stand out in particular, one in Austria, the other in Idaho. Researchers in both programs initially began their experiments with vegetable oils, but quickly discovered (after ruining quite a few test engines) that they were either going to have to modify the millions of diesel engines in use around the world to burn vegetable oil, or they were going to have to modify the vegetable oil. They decided on the second strategy. The modern fuel now known as biodiesel was largely the result of these two research programs. (For those who would like to learn more about the history of biodiesel, I recommend my 2005 book, *Biodiesel: Growing a New Energy Economy*.)

BIOFUELS 101

Ethanol and biodiesel have been receiving quite a lot of attention lately, especially in the U.S. media, which has suddenly realized that there's something exciting going on in the biofuels sector. But many people (even some in the media) are unclear about which biofuels are appropriate for which kinds of vehicles. Actually it's fairly simple: ethanol is for gasoline-powered vehicles; biodiesel is for diesel-powered vehicles. Ethanol can be mixed with or substituted for gasoline in spark-ignition engines (engines that have spark plugs, used to power most automobiles in the United States). Small percentages of ethanol blended with gasoline (10 percent or less) can be used in any gasoline engine, while larger percentages require a flexible fuel vehicle that can burn up to 85 percent ethanol, or E85 (some vehicles can even run on 100 percent ethanol).

Biodiesel, on the other hand, can be mixed with or substituted for diesel fuel in compression ignition (diesel) engines. The most popular biodiesel blend in the United States is called B20 (20 percent biodiesel, 80 percent petroleum diesel), but there are many other blends available in some locations. Diesel engines manufactured after 1994 can easily burn 100 percent biodiesel (B100) without any modification to the engine, although most engine manufacturers do not recommend this practice. Nevertheless, thousands of drivers use B100 regularly with no ill effects, except for some cold-weather issues, which I will describe later. Although there are substantial regulatory issues involved, both ethanol and biodiesel offer considerable opportunities for small-scale, local, cooperative production initiatives, especially at the small family farm or community level. The main advantages of using biofuels include their ability to be mixed with (or substituted directly for) fossil fuels, their compatibility with the existing distribution network, and their lower levels of harmful emissions compared with their fossil fuel equivalents.

But how can liquid fuels be made out of plant matter? Both biodiesel and ethanol can be made in various ways from different feedstocks, each having its own set of environmental impacts. The basic processes used are not especially complicated, and can actually be conducted as small-scale, backyard projects. In the case of ethanol, the process is almost as old as the hills, and has been avidly practiced by backwoods moonshiners for centuries. Most governments, however, have historically taken a dim view of their activities.

ETHANOL

Whether it's made legally or illegally, ethanol (ethyl alcohol, alcohol, or grain spirits) is made by converting the carbohydrate portion of biomass into sugar, which is then converted into ethanol in a fermentation process that is similar to brewing beer. This is the drinkable version of alcohol, and the active ingredient in beer, wine, and spirits (the nondrinkable version, known as methanol or wood alcohol, is poisonous). Simply stated, here is how the process works. The grain that is used as the feedstock is usually ground up to make the starches more available. The ground grain is mixed with water, cooked briefly, and enzymes are added to convert the starch to sugar using a chemical reaction called *hydrolysis*. The sugar is then fed to yeast, which digests the sugar and water to produce alcohol (ethanol) and carbon dioxide. The ethanol is separated from the mixture by distillation and the water is removed from the mixture using a dehydration process. The ethanol is then normally "denatured" with the addition of between 1 and 5 percent gasoline, making it unfit for human consumption, but usable as a fuel for internal combustion engines. Moonshiners would be appalled at this last step.

Ethanol offers a number of advantages over gasoline. Unlike gasoline, ethanol contains oxygen, which increases the efficiency of the combustion process, reducing tailpipe emissions of carbon monoxide (CO), nitrogen oxides (NO_x), particulate matter, and ozone-forming air toxins. Much of the octane needed for the performance of today's gasoline comes from highly toxic aromatic chemicals such as benzene, toulene, and xylene. Ethanol enhances fuel octane and generally provides a cleaner-burning alternative, although a few researchers have raised questions about increased ozone levels in warmer climates where low blends (5 to 10 percent) of ethanol with gasoline are used. Ethanol also has a lower energy content than gasoline, 83,333 Btu per gallon versus 124,800 Btu for gasoline. On the other hand, ethanol has a higher octane rating than gasoline (129 versus 91), making up somewhat for the reduced fuel economy. Ethanol also tends to clean out dirty deposits in fuel systems, sometimes requiring more frequent fuel filter changes, and acts as a gas-line antifreeze (or "dry gas") by absorbing moisture. Ethanol is biodegradable, with no lasting harmful effects on the environment if accidentally spilled.

Ethanol is the most widely used liquid biofuel today around the world, with about 12 billion gallons produced annually. Feedstock crops include sugarcane, corn, wheat, sugar beets, and a variety of other grains. The top five producers are the United States, Brazil, China, India, and France—which combined account for almost 90 percent of global production.[4] Since the mid 1970s, Brazil has had a deliberate government program to replace imported gasoline with domestically produced ethanol made from locally grown sugarcane. Originally the program was heavily subsidized, but those subsidies were ended in the mid 1990s, and even without them the price of ethanol today is only half that of gasoline. Ethanol currently accounts for 40 percent of the fuel sold in Brazil, and the nation intends to meet virtually all of the fuel needs of its automobile fleet with ethanol in the near future.[5] Not surprisingly, flexible-fuel vehicles that can run on gasoline, ethanol, or any mixture of the two have become extremely popular in Brazil, and seven out of every ten new cars sold in Brazil are flex-fuel.[6]

In 2005, U.S. production of ethanol was 3.9 billion gallons annually from 96 refineries based mainly on corn as a feedstock, according to the Department of Energy. But demand for ethanol is strong—and growing dramatically (as of August 2006, those figures had risen to 4.8 billion gallons annually from 101 refineries). The industry set a new production record in

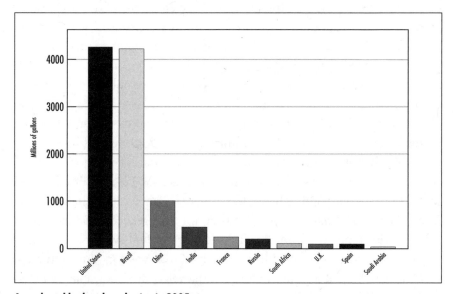

Annual world ethanol production in 2005. Based on figures from the Renewable Fuels Association.

January 2006 with 288,000 barrels per day (b/d), compared with 241,000 b/d the previous year, according to the U.S. Energy Information Administration. There are 41 plants and 7 expansions under construction that will soon add another 2.8 billion gallons to the annual production capacity.[7] Despite this dramatic growth, current ethanol production represents only 3 percent of the approximately 145 billion gallons of gasoline consumed in the United States annually. Ethanol has been blended with about 30 percent of the U.S. gasoline supply as an oxygenate or fuel extender for use in gasoline-powered vehicles. More recently, ethanol has been the choice of the oil industry to replace MTBE, a gasoline additive that was supposed to reduce air pollution but wound up polluting groundwater instead. This switch has spurred unprecedented demand for ethanol since early 2006, and the industry has been hard-pressed to meet that demand.

The dramatic growth in new ethanol plant construction, however, has a darker side. One of the most worrisome aspects is the appearance of new coal-fired ethanol refineries. That's right, coal-fired. This bizarre development even has some long-time ethanol supporters scratching their heads. In December 2005, a new ethanol refinery in Goldfield, Iowa, began producing what supporters describe as the "clean, renewable fuel of the future." The only problem is that this refinery uses three hundred tons of coal a day in the production process instead of natural gas. Another coal-fired ethanol plant is being constructed in the town of Nevada, Iowa, and at least three other refineries that will rely on coal are being built in Montana, North Dakota, and Minnesota. This trend, which is based on the lower price of coal compared to natural gas, has the potential to undermine the whole rationale for switching to ethanol in the first place, according to many observers, and is quickly becoming a public relations nightmare for an industry that doesn't need any more problems. "If your goal is to reduce costs, then coal is a good idea," says Robert Brown, director of Iowa State University's Office of Biorenewables. "If the goal is a renewable fuel, coal is a bad idea. When greenhouse-gas emissions go up, environmentalists take note. Then you've got a problem."[8]

In a related development, the huge surge in interest in ethanol has resulted in a dramatic shift in plant ownership patterns since 2005. In prior years well over 50 percent of ethanol plants were owned by farmers' cooperatives. But beginning in 2005, the number of farmer-owned plants plummeted to only six of the forty-two new projects under construction, according to the Iowa Renewable Fuels Association.[9] This dramatic shift has raised some concerns, especially among farmers, who view local community ownership of these plants as a bright spot in an otherwise difficult environment that has repeatedly challenged their survival skills in recent years. The real impetus for the recent surge has been the passage of the Energy Policy Act in August 2005, which set a target of 7.5 billion gallons of renewable fuels by 2012 and offered incentives to help reach that goal. The excellent financial returns on ethanol investment—often in the 20 to 30 percent range in recent years—has also finally attracted the attention of Wall Street. Not everyone is happy with this development. "Today the rise of giant plants and absentee plant ownership threatens to divorce our agricultural and even economic development goals from our goal of reducing dependence on imported oil," says David Morris from the Institute for Local Self-Reliance. "Something needs to be done. Now."[10]

Energy Balance

A modern ethanol plant in the United States can produce around 2.8 gallons of ethanol from a bushel of corn.[11] But there has been an ongoing and longstanding debate about ethanol's net energy balance—that is, how much energy needs to be expended in order to produce the energy contained in ethanol. If it takes more energy to produce a gallon of ethanol than the ethanol contains, the energy balance is negative (not good). If it takes less energy to produce the ethanol than the energy in the fuel, the balance is positive (good). With corn-based ethanol, the energy balance is just slightly positive (around 1.3 units out for every 1 unit in) according to most researchers (or negative according to a few), owing to the intensive nature of modern farming methods and the amount of energy required in the distillation process. This relationship can also be expressed as an *energy balance ratio* of 1.3 to 1. A newer type of ethanol, *cellulosic ethanol*, has a positive net energy balance, two to three times better than ethanol made from corn (more on cellulosic ethanol later).[12]

In an attempt to put this controversy to rest, several recent independent studies were conducted and then published in *Environmental Science & Technology* magazine and *Science* magazine. "What you see very clearly from both studies is that those [ethanol studies] that have found negative returns are decided outliers from a very large and solid set of alternative studies," says Nathaniel Greene, a senior policy analyst for the Natural Resources Defense Council (NRDC). "We need to move on now and start figuring out how we [can] use this technology and advance it to get as much out of it as possible."[13]

Nevertheless, there also have been serious lingering concerns raised by both scientists and environmentalists about the damage caused to farmland by large-scale agribusiness corn production, such as soil erosion, as well as pesticide and fertilizer runoffs that pollute waterways. These are legitimate concerns. But if peak oil or runaway global warming (or both) cause serious disruptions to large-scale agribusiness, then the questions about the disadvantages of corn as an ethanol feedstock may become moot. Scientific questions aside, ethanol does offer a number of obvious advantages. It is made from local, renewable sources; it helps reduce the amount of imported oil; and it keeps more money circulating in rural economies, especially if the ethanol plant is locally owned and operated.

Chippewa Valley Ethanol Cooperative
There are dozens of locally owned and operated ethanol plants in the United States. One prime example is located in the small city of Benson (population 3,237), located in west-central Minnesota. The idea for an ethanol plant in the Benson area originated from discussions between a local farmer and the manager of the local electric cooperative. They saw the new ethanol plant as a way to improve returns for area corn producers. The Chippewa Valley Agrafuels Cooperative was formed with over 650 shareholders made up of corn producers, grain elevators, and local investors. Today, that number stands at 950, and farmers and local community investors still make up the vast majority of the shareholders.

The cooperative then teamed up with an experienced designer-builder to form Chippewa Valley Ethanol Company, LLC (CVEC). Construction on the plant began in June 1995 and was completed in April 1996—ahead of schedule and well within budget. The first bushel of corn was ground on April 26, 1996. The plant averaged 98 percent of design capacity over the first six months of operation and produced 17 million gallons of ethanol during fiscal year 1997. The plant produced a record 42.9 million gallons

Chippewa Valley Ethanol Company, Benson, Minnesota. Chippewa Valley Agrafuels Cooperative.

in 2005. CVEC, which is a wholly owned subsidiary of the cooperative, now supplies E85 ethanol to over sixty service stations in Minnesota. In addition to fuel ethanol, CVEC also produces large quantities of industrial alcohol products.[14]

As noted in previous chapters, the state of Minnesota has been a leader in the support of renewable energy in general, and in 2005 approved a bill calling for 20 percent replacement of petroleum in its gasoline supply by 2012 through an E20 initiative. In the same session, the state also implemented the first statewide biodiesel requirement (B2) in the nation. These initiatives received strong bipartisan support. In an effort to stay abreast of the rapidly changing renewable energy markets, CVEC has decided to expand its production capacity further and integrate biomass conversion technologies at its Benson facility. The biomass gasification initiative will be a codevelopment project with Frontline BioEnergy of Ames, Iowa. The project will allow CVEC to gasify its own distillers grain by-product, as well as crop and wood residues to displace increasingly expensive natural gas for process heat in its ethanol production.

"We are surrounded by a sea of agricultural residue that goes unused, and at the same time we are looking at higher and higher natural gas prices," says CVEC general manager Bill Lee. "We are concerned about the price of natural gas long-term; it's our second largest input cost after corn. At the same time, we make a renewable fuel, and we'd like to move away from fossil inputs as much as possible. We still are unfortunately bedeviled by the whole energy balance question, and we see moving in this direction as one way to improve that balance. There are also processes that take synthesis gas directly to chemical products, and we envision that this is a route our company could take in the future. Down the road we like to think that we are putting some pieces in place that will allow us to participate in a cellulosic ethanol economy."[15]

The recent trend toward larger ethanol plants and less farmer ownership is a matter of considerable discussion—and concern—in Benson, according to Lee. "That's a very hot-button question out here these days," he says. "A few of the other farmer-owned plants have agreed to sell part or all of their operations to Big Money. The CVEC board and I see a huge difference between local ownership versus outside ownership, and we see the lion's share of the benefit of our operation coming from local ownership. Consequently, it would be highly inconsistent with our value set to allow substantial outside ownership." It remains to be seen in five or ten years how much local ownership of ethanol plants will remain. "I think we have reached the high-water mark, and have already come off that mark in terms of the percentage of the industry that is going to be owned by farmers," Lee continues. "It's kind of sad to see that. But on the other hand, we need more capital than can be raised in farm country, so new capital needs to come into the industry in order to grow it to a significant new scale. All we can do is control what we are doing here in our own community."

Several years ago, Lee became a member of the National Resources Defense Council, and has been active in trying to reach out to the environmental community, even to those in that community who have been hostile to corn-based ethanol. "I have been telling them that maybe corn ethanol isn't perfect, and maybe corn agronomy right now isn't the most sustainable practice, but it's better than imported petroleum," he says. "And by the way, the people who are making corn ethanol today are the ones who will be making cellulosic ethanol in the future, so stop beating us up; let's see if we can work together."[16]

Ethanol Limits

Although ethanol has come a long way, what the industry needs are more cars that can run on the fuel (and more filling stations that offer it) to make ethanol a more effective alternative nationwide, according to ethanol supporters. Others might point out that encouraging more cars isn't a solution; however, the biggest obstacle of all for ethanol is the limited availability of land on which the feedstock crops can be grown. Around the world, there is already intense competition between agriculture, forests, and urban sprawl, and any large-scale attempt to increase the amount of cropland for biofuel production will make the situation even worse.

Increasing the energy yield per acre is one useful strategy for making the best use of currently available cropland. In the United States, corn is the primary ethanol feedstock, but unfortunately it is also one of the least efficient. Ethanol yields per acre for sugar beets grown in France and sugarcane grown in Brazil are roughly double U.S. corn yields. However, sugarcane only grows in a few southern locations in the United States, and even sugar-based ethanol crop strategies have their limits. The dramatic growth in demand for sugar in the past year or so, especially in Brazil, due to the equally dramatic increase in demand for ethanol, has raised some concerns that we may be approaching "peak sugar" along with peak oil. Sugar production is already lagging behind demand, and the market is getting tight. Whether it's sugarcane, sugar beets, or corn, it is unrealistic to think that these crops can meet all our present demand for liquid fuels, and anyone who still thinks this is possible is in for a rude awakening. "We're beginning to run into a limit of how much ethanol we can get from corn," U.S. Energy Secretary Samuel Bodman conceded, at a meeting of the Greater Kansas City Chamber of Commerce on March 10, 2006. Bodman noted that the ethanol industry now consumes about 14 percent of the nation's total corn crop, and that research into alternative feedstocks is needed to avoid supply problems with corn.[17]

Cellulosic Ethanol

That research for alternative ethanol feedstocks has actually been going on for many years, and one of the main subjects of interest is *cellulosic* (cellulose-based) ethanol. This is the Holy Grail of the ethanol industry. Remarkably, this is no starry-eyed vision; the technology is more or less in hand, and commercialization is possible within the next few years or so. Scientists and researchers around the world are in a frenzied race to make this happen for what many describe as a "second-generation biofuel" (conventional ethanol and biodiesel being the first generation). The National Bioenergy Center of the National Renewable Energy Laboratory in Golden, Colorado, has been working with industry partners for years to help develop this technology. "The potential for cellulosic ethanol is huge," says George Douglas, a NREL spokesperson. "A recent study estimated that a third of our present transportation fuel needs could be met by cellulosic ethanol made from current biomass feedstocks. Combine that with increased vehicle fuel economy, and you are starting to make a real dent in petroleum use."[18]

The bulk of most plants is made up of cellulose, hemicellulose, and lignin.[19] The main idea is to make productive use of all of these materials, rather than just the seed, as in the production of corn-based ethanol. This can be accomplished by breaking down the cellulose and hemicellulose into component sugars so they can be fermented more easily to make ethanol. Producing ethanol from cellulosic biomass feedstocks rather than corn is seen as the best way to produce much larger quantities of ethanol from existing land resources without competing with food-crop and forest lands. In fact, the same land can actually produce both food and fuel under certain circumstances, providing farmers with additional income. These cellulosic feedstocks include agricultural and forest residues, grasses, and fast-growing trees—the same feedstocks I described in the last chapter for the many other potential uses of biomass—so there are clearly limits on how much can be diverted for ethanol production.

Although the process to make cellulosic ethanol is similar to traditional ethanol production using grain or corn, the method is more difficult and expensive because cellulose requires more complex pretreatment and other steps before the sugars can be fermented. However, a number of promising new technologies are being developed that use enzymes to break down cellulose to release the plant's sugars, which can then be fermented into ethanol.[20] Best of all, this technology can either be scaled up or down to match local requirements. The test facility at NREL only handles a ton of dry cellulosic feedstock a day, not a very large operation by most industrial standards. Think microbrewery, to get a sense of the size and scale of the facility. "We believe this technology is very, very close," Douglas says. "A lot of other people think so too. It's just a question of getting the cost down a little bit more to make it competitive."[21]

The world's first demonstration cellulosic ethanol plant is located in Ottawa, Canada, where Iogen Corporation has been brewing up test batches of the stuff for over two years. The plant, which has an annual production capacity of 260,000 gallons, has used straw from wheat, barley, oats, and rice, cornstalks, bagasse residue from sugarcane processing, and wood chips as feedstocks, according to Iogen president Brian Foody. "There's also research going on with energy crops like switchgrass," he says.[22] The company's cellulosic ethanol technology, which has been largely focused on the development of special enzymes that convert cellulose into sugar, is the result of over twenty-five years of research and development. Established in the 1970s, Iogen has become one of Canada's leading biotechnology firms, and is a manufacturer of enzyme products for use by the pulp and paper, textile, and animal-feed industries. It is also a world leader in technology to produce cellulosic ethanol.[23]

In 2002, Shell Global Solutions, the oil company's technology division, decided to make a $45 million investment in Iogen to speed development of a cost-effective process for making cellulosic ethanol. That investment has produced measurable results. Petro Canada took delivery of the first 1,300-gallon shipment from the plant in April 2004. Since then the facility, which has generally been operating at only about 10 percent of its capacity, consumes about thirty tons of straw per week.

Simply stated, here's how the process works. Large, one-thousand-pound bales of straw are loaded into a high-speed shredder, which reduces the straw to a light, fibrous fluff. After being shredded, the fiber goes through a "steam explosion" process that breaks up its structure even further. The fiber is then combined with water, heat, and enzymes in a large, sealed, stainless steel cylinder in a step called "enzymatic hydrolysis." The hot slurry sits in the cylinder for several days while the enzymes convert the fiber into sugar. The leftover woody matter, or lignin, from this process is dried and compressed into cakes that could be used as fuel in future cellulosic ethanol production facilities. The sugar is then processed into ethanol using the same general strategy mentioned earlier for making ethanol from grains.[24]

Although much work remains to be done, the Ottawa demonstration program has made significant breakthroughs, and on January 8, 2006, Iogen, Volkswagen, and Shell Oil announced that they were conducting a joint study to assess the economic feasibility of producing cellulosic ethanol in Germany on a commercial scale. "Iogen has demonstrated that clean, renewable fuels for transport are no longer a dream; they are a reality," Brian Foody said. "Today's announcement marks the first signal of what could be a major change coming in the European fuel market. It will show that by integrating vehicle and fuel technologies, we can meet the ambitious, but necessary challenge of reducing reliance upon fossil fuels."[25] Iogen is ahead of other companies who are trying to commercialize cellulosic ethanol, according to John Ashworth, an official of the U.S. Department of Energy's National Renewable Energy Laboratory which (as mentioned above) has a pilot plant for testing cellulosic ethanol concepts.[26]

Backyard Ethanol

But you don't need to be a multinational corporation to produce ethanol. Many people, especially farmers who have abundant feedstocks, produce ethanol for their own use. And it's legal. You do need to get a "small fuel producer" permit from the U.S. Bureau of Alcohol, Tobacco, and Firearms (www.ttb.gov/alcohol/permits.htm). The form and process are relatively simple. You will need to file an annual report on your production and how you disposed of it. Permits are issued for small, medium, and large production plants. Small groups wishing to produce ethanol can also do so (without having to go through the much more complicated process of obtaining a commercial permit) by forming a company for fuel production. You cannot sell the ethanol, even to each other, but the fuel belongs collectively to the company and can be used by its members. This sort of strategy could be particularly useful for small, locally owned, community-based fuel-production initiatives.

The equipment for making backyard ethanol is more complicated than that required for biodiesel, and the process is somewhat more involved. However, the only difference from the ethanol production process I mentioned earlier is that a backyard ethanol operation can be conducted on a very small scale. You will need a still, and you will also have to learn how to ferment beer. But this isn't rocket science; folks have been doing this for centuries under extremely primitive conditions. If you are relatively handy you can build your own still out of basic materials and then learn how to operate it yourself. One of the better books on this is *The Alcohol Fuel Handbook* by Lynn Ellen Doxon (see bibliography), where you will learn how to make ethanol from a wide range of feedstocks, including but not limited to apples, beans, buckwheat, corn, lentils, rice—even split peas. I can see it now in bold letters painted on the door, This Truck is Pea-Powered. Once you master the basic process, it's mostly a question of learning by observation and practice how to produce a successful batch of fuel. There are also vast resources for home-brew ethanol available on the Internet. One of the best places to start is at Journey to Forever (http://journeyto forever.org/ethanol.html), where you will find lots of general information, small-scale production guidance, many different forums, how-to books, and much, much more. Happy brewing.

BIODIESEL

Biodiesel is the second most widely used liquid biofuel today around the world (after ethanol), with about 800 million gallons (2.8 million metric tons) produced annually.[27] The top five producers are Germany, France, the United States, Italy, and Austria—which combined account for around 2.6 million metric tons (741 million gallons). Biodiesel has been manufactured on an industrial scale in Europe since 1992, and today Europe accounts for about 90 percent of world production. Germany is the undisputed global leader, producing about 1.6 million metric tons (456 million gallons) in 2005.[28] Although biodiesel has been around for many years in Europe, it is a relative newcomer in the United States, where until recently it was mainly used and promoted by Midwestern soybean farmers and a small group of backyard enthusiasts. But not any more. Biodiesel has suddenly emerged in the past few years as a mainstream vehicle and home heating fuel (or fuel additive). U.S. biodiesel production has vaulted from 500,000 gallons (1,754 metric tons) in 1999 to 75 million gallons (263,158 metric tons) in 2005. Production is projected to double to more than 150 million gallons by the end of 2006, and the industry is expected to continue to expand dramatically to meet the rapidly growing demand. Biodiesel has become America's fastest growing alternative fuel according to the Department of Energy.

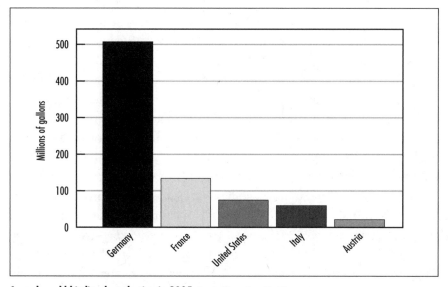

Annual world biodiesel production in 2005. Based on figures from F.O. Licht.

Until fairly recently, biodiesel was only available at a handful of obscure locations in the United States, but in 2001 the first two public biodiesel pumps were opened. Now, more than one thousand filling stations offer biodiesel to the public and over fifteen hundred petroleum distributors carry it nationwide. In addition, more than six hundred fleets use biodiesel, including government and military, commercial, and school bus fleets, to name just a few.[29] Despite this dramatic growth, 75 million gallons is a drop in the bucket compared to the approximately 58 *billion* gallons of so-called middle-distillate fuels (including diesel fuel, heating oil, kerosene, jet fuels, and gas turbine engine fuels) currently consumed in the United States every year. Of that total, about 40 billion gallons is for on-highway vehicle use, while the rest is divided among a wide range of residential, commercial, and industrial purposes.

Biodiesel 101

Biodiesel, a diverse group of diesel-like fuels, can be easily made through a simple chemical process known as *transesterification* from virtually any vegetable oil, including (but not limited to) soy, corn, rapeseed (canola), cottonseed, peanut, sunflower, mustard seed, and hemp. But biodiesel can also be made from recycled cooking oil (referred to as "yellow grease" in the rendering industry) or animal fats. One Vietnamese catfish processor is even using fish fat as a biofuel feedstock.[30] There have even been some promising experiments with the use of algae as a biodiesel feedstock. As long as the resulting fuel meets the American Society for Testing and Materials (ASTM) biodiesel standard (D-6751), it's considered biodiesel in the United States, regardless of the feedstock used in its manufacture (in Europe, the standard is EN 14214). And the process is so simple that biodiesel can be made by virtually anyone, although the chemicals required (usually lye and methanol) are hazardous, and need to be handled with extreme caution.

Simply stated, here is how biodiesel is made. The transesterification process is initiated by adding carefully measured amounts of alcohol (methanol) mixed with a catalyst (sodium hydroxide—lye—the same chemical used to unclog kitchen or bathroom drains) to the vegetable oil. The mixture is stirred or agitated (and sometimes heated) for a specific length of time. If used cooking oil is the feedstock, the process requires a bit more testing, lye, and filtration, but is otherwise essentially the same. During the mixing, the oil molecules are split or "cracked" and the methyl esters (biodiesel) rise to the top of the settling/mixing tank, while the glycerin and catalyst settle to the bottom. After about eight hours, the glycerin and catalyst are drawn off the bottom, leaving biodiesel in the tank. The whole idea of the process is to remove the thick, sticky glycerin from the vegetable oil, so the remaining biodiesel will flow easily and combust properly in a modern diesel engine without leaving damaging deposits inside the engine.

In most cases the biodiesel needs to be washed with water to remove any remaining traces of alcohol, catalyst, and glycerin. In this procedure, water is mixed with the biodiesel, allowed to settle out for several days, and then removed. The wash process can be repeated if needed, but it is time-consuming. Not everyone agrees on whether the water wash is necessary. A few smaller producers who are making biodiesel for themselves skip the process, while commercial producers usually must do it to meet industry standards. In the case of some larger, more sophisticated manufacturing facilities, the transesterification process itself is so carefully controlled and refined that the water wash is not needed. There are, of course, quite a few technical variations on this entire process for large-scale industrial operations, but the general transesterification procedure is similar.[31]

As the amount of biodiesel being produced grows exponentially, the quantities of glycerin by-product grows apace. Glycerin has always been a niche market that is highly sensitive to oversupply, and the recent exponential growth of this commodity as a result of biodiesel production has caused the world glycerin market to collapse. As a result, traditional glycerin manufacturing plants around the world have been closing, while new ones that use glycerin as feedstocks for epoxy resins, propylene glycol, and other products have been opening. Recently, glycerin has even been used by one California company, InnovaTek Inc., as a source for the production of hydrogen.[32] Trying to develop new uses for glycerin has been keeping a lot of people awake at night.

Biodiesel Characteristics

Biodiesel offers a number of advantages when compared with diesel fuel. First, biodiesel can be used in any modern diesel engine without any modifications to the engine. Biodiesel has excellent lubricating properties and will lubricate many moving parts in the engine, increasing engine life. It also has a higher *cetane number* than petrodiesel (cetane is a measure of the ignition quality of the fuel), indicating better ignition properties. On the downside, the energy content of biodiesel is about 10 to 12 percent lower than that of petrodiesel (about 121,000 Btu per gallon compared to 135,000 Btu for diesel fuel). Biodiesel has an oxygen content of around 10 percent, more than that of petrodiesel fuel. The oxygen results in more favorable emission levels, but also results in reduced energy content. Although this causes a slight reduction in engine performance, the loss is partly offset by a 7 percent average increase in combustion efficiency. Generally speaking, the use of biodiesel results in about a 5 percent decrease in torque, power, and fuel efficiency in diesel engines.[33] However, most people can't detect any noticeable difference in the performance of their vehicles.

Biodiesel is free of lead, contains virtually no sulfur or aromatics (toxic compounds such as benzene, toluene, and xylene), and results in substantial reductions in the release of unburned hydrocarbons, carbon monoxide, and particulate matter (soot), which has been linked to respiratory disease, cancer, and other adverse health effects. The production and use of biodiesel results in a 78 percent reduction in carbon dioxide emissions, according to a joint U.S. Department of Energy and U.S. Department of Agriculture study published in 1998.[34] This is mainly because the plants used to grow biodiesel feedstock absorb most of the CO_2 emissions from biodiesel combustion. On the downside, emissions of nitrogen oxides (NO_x, a contributing factor in the formation of smog and ozone) were thought to be slightly greater with biodiesel. However, a recent study by the National Renewable Energy Laboratory concluded that vehicles using B20 do not produce an increase in NO_x emissions. Finally, biodiesel replaces the typically noxious exhaust smell of petrodiesel with an odor faintly like that of french fries or doughnuts, especially if it has been made from recycled cooking oil.

Operational Issues

Biodiesel does suffer from a few operational problems. Diesel engines are notoriously hard to start in cold weather. The reason for this is because petrodiesel begins to "cloud" at about 20°F (–7°C), the temperature at which paraffin wax crystals begin to form. Cloudy diesel fuel can clog fuel filters and keep the engine from starting or cause it to stall. When temperatures fall further, diesel fuel reaches its *pour point* (the temperature below which it will not pour). At this point diesel fuel generally stops flowing through fuel lines and diesel engines stop running. When the temperature drops even further, to about 15°F (–9.5°C), diesel fuel reaches its *gel point*, and becomes the consistency of petroleum jelly. Biodiesel, unfortunately, suffers from all of these problems, but at higher temperatures, making the situation even worse. The actual cloud point for biodiesel varies depending on what was used for the feedstock. Biodiesel made from used cooking oil or animal fats will cloud at higher temperatures than biodiesel made from virgin rapeseed/canola oil. But there are even different cloud points among various types of virgin oils. For example, biodiesel made from palm oil generally does not perform well in cold climates.

During the winter months in cold regions, petrodiesel fuel is normally altered with special winter formulations to help it perform better. There are also winterizing agents, antigel formulas, and other additives that can lower the cloud point of petrodiesel. These same agents and formulas can be added to biodiesel blends to improve winter performance as well (although the additives tend to work primarily on the petrodiesel portion of the mix). Another widely used strategy is to use a lower biodiesel blend, such as B20 or less, during the winter and a higher biodiesel concentration, such as B50 or B100, during the warmer months. For hard-core biodiesel purists who live in cold climates, special cold-weather heating kits are available for most diesel cars and trucks—although, generally speaking, the use of high biodiesel concentrations during the winter can be an invitation for trouble.

Another characteristic of biodiesel that can cause some problems, especially in older diesel engines, are its high solvent properties. On engines manufactured before 1994, the rubber seals, hoses, and gaskets will be degraded with the use of high concentrations of biodiesel, especially B100. If these rubber parts are replaced with biodiesel-resistant materials such as Viton or Teflon, B100 can be used without any problems. But biodiesel's solvent properties can also cause problems with old fuel tanks and fuel lines, which are typically coated with sludge. The biodiesel dissolves the sludge, which then ends up in the fuel filter, causing the engine to malfunction. However, once the sludge has been cleaned out of the fuel system, the engine should run without further trouble on biodiesel.

In warm climates, diesel engines, and especially their fuel tanks, are susceptible to bacteria growth, which can clog the fuel system and cause engine failure. This problem can occur with either petrodiesel or biodiesel fuels. The typical greenish to black bacterial slime grows in the absence of light but in the presence of moisture in the fuel tank as it feeds on the hydrocarbons in the fuel. The bacteria can be eliminated with the use of biocides, which are widely available at automotive parts stores, fuel dealers, and other retail outlets. Keeping the fuel tank as full as possible will minimize the amount of condensation, oxygen, and bacteriological activity.

Biodiesel's Uses
Over-the-road vehicles account for the vast majority of diesel fuel use in the United States. When most Americans think of transportation they generally think first of automobiles. But the transport sector extends well beyond cars to include trucks, buses, trains, boats, and planes. And the vast majority of these other types of transport are powered by diesel engines or, in the case of commercial aircraft, on diesel-like aviation fuels. It's no exaggeration to say that the vast majority of the world's heavy transport sector is diesel powered. This offers obvious potential for a wide range of uses for biodiesel.

Diesel-powered automobiles in the United States represent only about 3 percent of the total fleet (as opposed to around 50 percent in many European countries), so biodiesel has not had much effect on that sector so far. But many fleets, especially mass transit systems, have significant numbers of diesel-powered vehicles; as noted above, over six hundred fleets now use biodiesel blends to power their buses, trucks, and other heavy equipment. Biodiesel is also being used to fuel larger numbers of diesel-powered farm vehicles, machinery, and other equipment, and a growing number of ferries and other maritime interests are using biodiesel blends. A small number of railroads are experimenting with biodiesel-powered trains as well.

Bioheat
In the United States, until fairly recently, biodiesel has been promoted mainly as a fuel for diesel-powered vehicles. But many people (even some in the biodiesel community) don't realize that biodiesel can also be used as a heating-fuel additive or replacement in a standard oil-fired furnace or boiler. That's because number 2 heating oil is virtually the same as standard petrodiesel vehicle fuel, and biodiesel can be mixed at any percentage with number 2 oil. When used for space heating, biodiesel is sometimes referred to as "biofuel" or "bioheat" in the United States.

The conversion process for an oil-fired furnace or boiler is just as simple as the conversion for a diesel engine—just add the biodiesel to the fuel tank. No new heating appliance or expensive retrofitting is required as long as the blend is B20 or less (for higher concentrations of biodiesel, a small rubber seal in the fuel pump needs to be replaced with a synthetic seal made out of Viton or Teflon). What's more, since it's located in a protected space, using biodiesel as a home heating fuel has fewer of the cold-weather operating problems that are associated with using biodiesel in vehicles during the winter. Bioheat also reduces harmful emissions and burning it helps to keep the boiler or furnace cleaner as well. Using biodiesel as a heating fuel is such a simple idea you have to wonder why nobody thought of it sooner. Actually, someone did, because biodiesel has been used as a heating fuel in Italy, France, and a number of other European countries for many years. Now the United States is beginning to catch up. Bioheat is now available from a rapidly increasing number of fuel distributors, especially in the Northeast.

Feedstock Issues

In the United States, soybeans are the feedstock of choice for making biodiesel, not because soybeans yield a lot of oil (relatively speaking, they don't), but because Midwestern soybean farmers have promoted biodiesel for many years as an alternative market for historic oversupplies of soybean oil. One bushel of soybeans is needed to produce about 1.5 gallons of biodiesel. A lot of research is ongoing to find higher-yielding biodiesel feedstock crops that will grow in various North American climate zones. Canola (known as rapeseed in Europe) and mustard are the focus of much of that research.

Biodiesel's energy balance is far better than (corn) ethanol's, with about two to three times more energy contained in the fuel than the energy needed to produce it (a small number of researchers—most notably David Pimental and Tad Patzek—dispute this, but they definitely are in the minority). When field crops are used as the feedstock, biodiesel, like ethanol, relies mainly on a large-scale, oil-based, agribusiness model of production, which can damage soils and cause water pollution. The use of genetically modified crops as feedstocks is another troubling issue for many in the environmental community. And even under the most optimistic scenarios, biodiesel probably will only be able to meet about 15 to 20 percent of our present diesel fuel demand given current technology (though these percentages could be much higher with the use of algae as a feedstock). And since biodiesel and ethanol generally compete for the same croplands, dramatic increases in production for one fuel tends to limit growth in the other.

In the European Union, the upper limits of field-crop production of biodiesel feedstock crops (mainly rapeseed) are beginning to be reached on so called set-aside land,[35] resulting in a significant shift to overseas sources. Unfortunately, some of these overseas strategies rely on the conversion of forestland to large-scale plantations of biofuel crops, resulting in serious degradation of soils, to say nothing of the loss of the forests. This recent, and growing, trend in large-scale "outsourcing" of biofuel feedstock production to Third World nations—some of which can barely feed their own populations—is beginning to raise serious concerns in the environmental community, which would otherwise generally support the shift from petroleum to biofuels. These concerns are not limited to the environmental community. Even Royal Dutch Shell, the world's top marketer of biofuels, is mindful of this issue and considers using food crops to make biofuels to be "morally inappropriate" as long as there are people starving, according to Eric Holthusen, the company's Asia/Pacific Fuels Technology Manager.[36] Some sort of sustainable biofuel feedstock production certification needs to be developed and implemented for biodiesel, especially from these Third World countries. The sooner the better.

Beyond field-crop feedstock production lies the Holy Grail of the biodiesel industry: algae. The use of algae as a biodiesel feedstock has received a lot of attention over the years and was the subject of intensive research by the National Renewable Energy Laboratory from 1976 to 1998. The main focus of the program was the production of biodiesel from algae grown in ponds utilizing waste CO_2 from coal-fired power plants, and considerable progress was made. Although the NREL program was ended because of budget cuts, the researchers hoped that their work would be used as a foundation for the future production of biodiesel from algae.

Those hopes have at least partly been realized, and there are a lot of initiatives currently underway around the world to move the results of the research into practical, commercial-scale applications. One such initiative is being pursued by GreenFuel Technologies, in Cambridge, Massachusetts. GreenFuel has followed the lead of the NREL research and is conducting a field trial at a 1,000-megawatt power plant owned by a major southwestern power company. The algae is harvested daily, and the oil that is squeezed from it is converted into biodiesel. In 2007, GreenFuel anticipates a number of additional demonstration projects, eventually scaling up to a full production system by 2009.[37] Some enthusiastic algae proponents maintain that much, if not all, of our liquid fuel needs could be met by algae, and a broad range of research activity continues. Expect to hear a lot more about this in the near future.

Biodiesel Communities

Because making biodiesel is relatively simple and can be very low-tech (an old 55-gallon drum is often used as the settling/mixing tank), it has attracted an enthusiastic community of backyard enthusiasts or "home brewers" around the world. In the United States, these folks generally refer to themselves as "the B100 Community," because they prefer to run their vehicles on pure (B100) biodiesel, rather than the lower-percentage blends advocated by the mainstream biodiesel industry. Although the mainstream industry and the B100 Community both make biodiesel, their focus is quite different, and they generally work with constituencies that tend to be polar opposites. Because it's a grassroots movement, the B100 Community has a very small-scale, local focus, while the mainstream industry is centered on a large-scale, corporate, agribusiness model. For years, there has been a good deal of tension between the mainstream industry, as represented by the National Biodiesel Board, and the B100 Community over a wide range of issues. But recently, some progress toward accommodation (or at least a truce) between these very different segments of the industry has been made. The B100 Community is very active around the country at the grass-roots level educating the general public about the many benefits of biodiesel and greater energy independence. The best description of the B100 Community can be found in Lyle Estill's excellent and entertaining book, *Biodiesel Power*. For those who want to make their own biodiesel, *From the Fryer to the Fuel Tank: The Complete Guide to Using Vegetable Oil as an Alternative Fuel* by Joshua Tickell is a good place to start. A more recent publication called the *Biodiesel Homebrew Guide* goes into more accurate detail about making biodiesel (and a safe, inexpensive biodiesel processor). It is available online from Maria "Mark" Alovert. (For more on all of these references, see the bibliography.)

Producers of biodiesel fall into three main categories: large-scale com-mercial, small-scale, and individual. Large-scale commercial producers can be corporate- or community-owned (often through local farmers' coopera-tives), and this latter group is where most of the current biodiesel action is taking place in many countries. Because of their size, and the fact that they sell huge quantities of fuel to government agencies, institutions, businesses, and the general public, large producers are subject to a wide range of fed-eral, state, and local regulations. The effort and considerable expense asso-ciated with meeting these regulatory requirements are simply viewed as part of the cost of doing business. At the other end of the spectrum, there are also quite a few individual "home brew" producers around the world, who tend to operate below the radar of most government oversight. It's hard to say exactly how many people are involved, but there is a lot of activity at the grassroots level. The small-scale producer, however, falls somewhere in between these two extremes, and in the United States at least, this leaves them in a difficult position. These days, a small producer probably makes somewhere between 100,000 to 500,000 gallons of biodiesel per year. Too large to escape detection by government, but too small to be able to easily afford the high cost of meeting regulatory requirements for selling ASTM-compliant fuel to the general public, small producers face formidable obstacles. In addition, virtually all government incentive programs and tax breaks favor large, corporate players, leaving the small producer essentially shut out. There is one strategy, however, that eliminates some, but not all, of these problems: the small cooperative.

Cooperatives

There are essentially two very different types of biodiesel cooperatives. The large, often farmer-owned cooperatives that are located in the Midwest, tend to be dominated by large-scale soy-farming interests, and represent the primary original political base for the biodiesel industry in the United States. West Central Cooperative in Ralston, Iowa, with its 3,500-plus farmer members, is a good example of a large and successful co-op. West Central's 12-million-gallon capacity biodiesel plant processes 90 million pounds of soybean oil into biodiesel and other products every year. Because of their size and volume of production, these large-scale cooperatives can afford to join the National Biodiesel Board, and through that membership gain access to the health-effects testing data required for EPA approval—a prerequisite for any large-scale fuel producer in the United States. Farmer-based biodiesel cooperatives have some additional potential in farming areas beyond the Midwest, and there has been some recent activity in other states outside of this region.

At the opposite end of the spectrum are the very small, nonfarmer-based, backyard cooperatives represented by portions of the B100 Community. This community-based model has some promise for modest success at the local level for small producers. Since cooperatives don't sell the fuel they make to the general public, they generally don't trigger some of the worst regulatory hurdles—like EPA approval—that stop most other small producers dead in their tracks. But this strategy has significant limitations. By all accounts, there are quite a few co-ops of various types in existence (or springing up) around the country, but not too many of them have actually produced much biodiesel. However, they do seem to spend a lot of time and energy democrat-ically discussing, debating, and planning on making biodiesel. And because there are many different ways of achieving that end, there are many, many variations on the small co-op theme. Nevertheless, small biodiesel co-ops generally seem to fall into one of three main categories:

- very small-scale producer,
- bulk buyer,
- large volunteer producer group.

The first category, very small-scale co-op producer, usually involves per-haps two experienced individuals who take care of all fuel-making chores for themselves and a small group of other subscribers. This model, which offers relative safety, quality control, and ease of management, has been fairly successful—as long as it stays small, according to Maria "Mark" Alovert of San Rafael, California, one of the B100 Community's most experienced and knowledgeable activists. Alovert has been involved in a number of co-ops over the years, so she knows what she is talking about.

The second category, the bulk-buy co-op, eliminates production chores completely (as well as possible production accidents) and normally offers less-expensive fuel to subscribers. Some bulk-buy co-ops use a large central storage tank where loads are delivered and where subscribers pick up their fuel. Other co-ops use distributed smaller storage tanks or barrels in mul-tiple locations. Regardless of the storage strategy, a bulk-buy co-op still requires quite a lot of coordination and oversight to work smoothly.

Finally, there is the kind of co-op where everybody gets trained to make fuel on a volunteer basis. Of all the models, this one potentially presents the greatest challenges, according to people who have been involved with them. This is especially true for groups of more than five fuel makers. "If you have any thoughts whatsoever that involve doing this with more than five people at a single site, think twice," warns Alovert.

The logistics involved with trying to train, maintain, and coordinate a large group of people to make consistently high-quality biodiesel safely are daunting. The Berkeley Biodiesel Collective in California (a producers' co-op), founded in September 2002, tried and tried to make this model work, but ultimately failed. After a thorough reorganization of its efforts, the collective did eventually find a successful model that involved a major training initiative and a decentralized approach involving very small groups combined with a bulk-buyer's club for the purchase of commercial biodiesel. "The shakeup and restructuring took a year to happen, during which a lot of people's time was evaporated by the misguided attempt at 'empowering people through democratic group fuelmaking' and similar nonsense that didn't take enough reality into account," Alovert says.[38]

Nevertheless, there have been a few other examples of successful small co-ops, such as the Co-op Power (220 members) biodiesel initiative in Massachusetts. Although the project has experienced numerous delays, Co-op Power hopes to break ground on its new Northeast Biodiesel plant in Greenfield in late 2006. The facility will use waste cooking oil as its main feedstock and is expected to produce 5 million gallons in its first year. But one of the most successful biodiesel cooperatives has been Piedmont Biofuels.

Piedmont Biofuels is a member-owned cooperative located in Pittsboro, North Carolina. It currently has about 226 members. The group has been leading the grassroots sustainability movement in North Carolina by using and encouraging the use of clean, renewable biofuels. Members are entitled to buy biodiesel from the co-op or to learn how to make their own using co-op equipment. The co-op has five retail outlets for biodiesel and is actively working at establishing more. Piedmont Biofuels also has a B100 fuel terminal (the first of its kind in the state), as well as a 1,600-gallon fuel delivery truck, which it uses to deliver biodiesel in the region. Most of the biodiesel the co-op sells is for "on-road" uses, meaning that the price of the fuel includes all state and federal taxes. A smaller amount of biodiesel is sold for "on-farm," home heating, marine, generator, and other "off-road" uses without road taxes. When sold for "off-road" purposes, the fuel is dyed to indicate that road taxes have not been paid. Regardless of its end use, all of the biodiesel sold by the co-op meets the ASTM (D-6751) quality specification.

The co-op also sells biodiesel reactors (the equipment in which the biodiesel is made). Local, micro-scale biodiesel production and consumption makes a lot of sense in the co-op's view, and it is committed to spreading the knowledge and technology necessary to produce quality biodiesel in small quantities (under 250,000 gallons per year). Piedmont

consults, designs, and builds small biodiesel reactors, and trains their cus-
tomers on how to use them properly. The first reactor the co-op sold went
to North Carolina State University, as part of the Solar Center's
Alternative Fuel Garage. Since then, Piedmont has designed and built a
number of units for its own use, along with a mobile reactor for the U.S.
Department of Energy through the State Energy Office and Central
Carolina Community College. The co-op also delivered one reactor to the
North Carolina Zoo, which was mounted on a trailer.[39]

Piedmont has a strong educational program. It created and runs the
Biofuels Program at Central Carolina Community College. Co-op members have taught, demonstrated, and lectured around the country—from
Solar Energy International in Colorado to North Carolina State to the
University of Florida. The co-op holds periodic workshops and confer-
ences on everything from reactor design and building to straight vegetable-oil conversions to co-op formation. "This is by far the biggest thing
that we do," says Lyle Estill, one of Piedmont's cofounders. "More time,
money, and effort is spent on education than anything else, but it's

Piedmont Biofuels members. Kent Corley.

extremely important. The number of gallons of fuel that we make is really insignificant compared to the number of people that go away thinking about energy in a different way."[40]

And, of course, Piedmont has been making biodiesel at its backyard facility in Moncure. People who want to learn how to make biodiesel out of waste vegetable oil can join the co-op and then join in with an experienced home-brewing crew. Home brewing at the co-op is a voluntary activity that rises and falls with the enthusiasm of the membership. It's legal to make your own fuel and use it to drive on public highways. In order to stay legal, however, the co-op pays road taxes on all homemade fuel consumed on behalf of its members. Dealing with regulations has been one of the largest challenges the co-op has faced. "We're out there speaking, blogging, writing books, shooting our mouths off, and as a result, we're just way too public to try to do anything under the table," Estill says. "We've been visited by the IRS three times, North Carolina Revenue twice. When we plug into the regulatory framework, what happens is the town makes a lateral pass to the county, which then passes to the state, the state passes to the feds, and we end up in this regulatory no-man's-land. Consequently, the regulator sitting on the other side of the table doesn't know what to do. It drives us crazy."

Despite the many hurdles, Piedmont Biofuels has just made the transition from being a small, backyard cooperative biodiesel producer to being a small cooperative industrial producer with its new Piedmont Biofuels Industrial LLC venture. "After years of successfully resisting the urge to go into commercial biodiesel production, we have finally succumbed," Estill admits. Located in an abandoned chemical plant at the edge of Pittsboro, the multifeedstock batch-process facility is designed to produce about a million gallons of biodiesel annually. "We are attempting to collect our feedstocks from a 100-mile radius and we're going to distribute our fuel within that area," Estill explains. "In our view, it's infinitely preferable to have 100 separate million-gallon plants scattered around the country in small towns rather than a single 100-million-gallon plant. We're not interested in being the next big fuel monopoly; we're just trying to help our town to meet its own needs." This model, with local feedstocks and local customers, makes a lot of sense for small towns everywhere.

By virtually anybody's standards, Piedmont Biofuels has been a remarkable success. Trying to replicate that success, however, is not as easy as one might assume. The national landscape is littered with "the remnants of biodiesel co-ops," according to Estill. "The main thing that we have done that many others have failed to do is survive," he says. "No one got mad and took their reactor and went home. No one fled when the IRS came to town. Leif [Forer] and Rachel [Burton] and I have stuck together, and that's one of our greatest strengths. We're the garage band, and we haven't broken up yet. I joke that what you need for success with a biodiesel co-op is easy. Start with a magnetic, female diesel mechanic . . . and then I always qualify it by saying if you want to be like Piedmont Biofuels, be careful what you wish for. I think we're probably going to have a board meeting every single Sunday this month; it's a lot of hard work."[41]

Biomass-to-Liquid

As I mentioned in chapter 5, biomass can also be gasified to produce a synthesis gas composed primarily of hydrogen and carbon monoxide, also called syngas, wood gas, biogas, biosyngas, or producer gas. The hydrogen can be recovered from this syngas, or it can be catalytically converted to methanol. The hydrogen can also be converted using a Fischer-Tropsch (FT) process into a liquid with properties similar to diesel fuel, known as Fischer-Tropsch diesel. This combination of biomass gasification and FT synthesis is a highly promising strategy for producing liquid transportation fuels. The process is sometimes called biomass-to-liquid, or BTL. Since it is based on biomass rather than coal or natural gas, this variant on the Fischer-Tropsch process is renewable.

Traditional biofuels such as biodiesel and ethanol are so-called "first-generation" biofuels that are made using the same parts of plants (seed, grain, or sugarcane crops) that are also used in food production. Competition between fuel and food for these crops has the potential to negatively affect both the availability and price of these crops. In contrast, biomass-based BTL fuel and ethanol produced from cellulose (cellulosic ethanol) are now being referred to as "second-generation" biofuels. These fuels are made by converting the parts of plants not used for food production. In this scenario, farmers are able to meet the needs of both the food and fuels industry from the same land, significantly increasing yields per acre while simultaneously improving income from that land.

Without getting too technical, here is how the process works. During the first stage of the three-step process, the biomass is broken down into a gas containing tar (volatile parts) and solid carbon (char) through (relatively) low-temperature (400°C or 752°F) heating in a manner similar to the syngas process described in chapter 5. During the second stage of the process, the gas-containing tar is combusted with oxygen at about 1,400°C (2,552°F) to turn it into a hot gasification medium. During the third stage of the process, the char is ground down into pulverized fuel and is blown into the hot gasification medium. The pulverized fuel and the gasification medium react in the gasification reactor and are converted into a raw synthesis gas. Once this gas has been treated to remove impurities, it can be used for producing Fischer-Tropsch diesel in a process where the reactive parts of the synthesis gas interact with a catalyst to form the hydrocarbons that make up BTL diesel.[42]

BTL diesel is colorless, odorless, and relatively low in toxicity. BTL diesel has virtually all of the same advantages and other characteristics as regular biodiesel, and there are no significant differences in performance. The renewable fuel's higher cetane number should result in improved combustion. In addition, BTL diesel is virtually interchangeable with conventional petrodiesel fuels and can be blended with them at any percentage. BTL fuels offer important emissions benefits compared with petrodiesel, reducing nitrogen oxide, carbon monoxide, and particulate matter. Like regular biodiesel, BTL diesel can also be stored and transported in the existing fuel-distribution network and can be used for fuel in regular diesel engines without modification.

On the downside, BTL diesel costs about 10 percent more than conventional diesel, and has a lower energy content than petrodiesel, which may result in slightly lower fuel economy and power.[43] What's more, some environmentalists have claimed that CO_2 emissions of the Fischer-Tropsch process exceed those for emissions of crude-oil refining. However, since the plants grown for the next feedstock crop absorb most of that CO_2, BTL diesel is much better from a greenhouse gas perspective than any other synthetic or conventional petroleum fuel or biofuel. Greenhouse gas emissions from BTL fuel are less than 10 percent of those from fossil fuels. The U.S. Department of Energy has also raised some concerns, noting that the FT process is energy-intensive, especially when compared to a similar process that uses natural gas as a feedstock. Of course, using natural gas as a feedstock has its own problems.

Finally, there is one other BTL fuel that deserves a few words, *dimethyl ether* (DME). A widely used propellant for aerosol products, DME is gaseous at ambient temperatures, but it can be liquefied under pressure and has some potential as an alternative vehicle fuel. Although it is usually made from coal or natural gas, it can also be synthesized from biomass-derived syngas. With a high cetane number, DME is clean-burning, sulfur-free, and has extremely low particulates. In Shanghai, China, a chemical company has started producing DME as an alternative fuel for diesel-powered transit buses in that city. Unfortunately, unlike BTL diesel, dimethyl ether requires special fuel-handling and engine-injection systems.[44] It is unclear at this time how useful DME will be in the future.

Although BTL diesel is being developed in various countries, it is the most advanced in Germany. Encouraged and supported by DaimlerChrysler and Volkswagen, CHOREN Industries of Freiberg, Germany, has developed a process that in some ways mimics nature's evolutionary process for producing crude oil—but on a much shorter time scale. CHOREN's small experimental (Alpha) plant has been operating in Freiberg for several years, producing a few hundred liters per day. A larger (Beta) plant under construction at Freiberg will be able to produce 16.5 million liters (4.4 million gallons) annually. The Beta plant, incorporating CHOREN's proprietary Carbo-V process along with additional technology from Shell, is expected to come online in early 2007. But that's not all. The first commercial-scale plant with a 200,000 metric ton (57 million gallon) annual capacity is planned for Lubmin near Greifswald in northeastern Germany in 2008–09. It is estimated that around 750 jobs will be created as a result of the Lubmin project.[45] In 2003, CHOREN was the first company to demonstrably produce BTL fuel outside of a laboratory environment. The company's unique, patented gasification technology is able to convert a wide array of biomass feedstocks such as wood chips, straw, or energy plants into a tar-free synthetic gas that may be converted into liquid transport fuel using FT synthesis (or alternatively for power generation or process heat).

In August 2005, Shell Deutschland Oil GmbH acquired a minority equity stake in CHOREN. This set the stage for the construction of the commercial BTL facility, and allows Shell to bring its expertise in middle distillate synthesis technology to the project. This technology, originally developed by Shell for the conversion of natural gas into synthetic oil products, can also be used to convert the biomass-based syngas produced by CHOREN into BTL diesel fuel. "We are particularly pleased to have won over a partner with Shell's FT knowledge and clean fuel experience," says Tom Blades, CHOREN's CEO. "It affirms our vision for the realization of large-scale production of BTL-derived transport fuels that meet the needs of today's and tomorrow's mobile society but without further burdening the environment."[46]

After combining its efforts with Shell in Germany, CHOREN has begun to evaluate the possibility of bringing its advanced gasifier technology to the United States. The company has been looking for potential BTL sites that can provide 1 million tons of sustainable biomass per year and that have the other necessary infrastructure and regulatory support for a 70 million gallon facility.[47]

STRAIGHT VEG

Most mainstream literature about biofuels doesn't even mention straight vegetable oil (SVO), except as a feedstock for biodiesel. But no discussion of biofuels in a post-peak-oil environment would be complete without at least a few words about one of Rudolf Diesel's original fuels. This strategy, referred to by many enthusiasts as "straight veg," makes use of waste vegetable oil (WVO), also known as yellow grease, collected from restaurants to fuel diesel engines *directly* without first going through the transesterification process to make biodiesel. This bit of alchemy is accomplished by adapting the vehicle to the fuel with a conversion kit, costing roughly $300 to $1,500 in the United States.

Basically, the typical conversion kit involves adding a parallel fuel system that consists of a second fuel tank, a heater, an extra fuel line to the engine, a filter, and a control that allows the driver to switch back and forth between the two systems. The vehicle has to be started with diesel fuel (or biodiesel) from the first tank, and then, after the engine heats up and thoroughly warms the WVO in the second tank to reduce its viscosity (thickness), the driver manually switches to the second tank (some systems feature an automatic control for this). At this point the vehicle is running on straight vegetable oil. When the vehicle needs to be shut off, the process is reversed. The driver manually switches back to tank number one (containing the diesel fuel) for a few minutes to ensure that all the WVO has been purged from the fuel line and engine. Failure to do so will cause serious problems the next time you try to start the vehicle.

While this strategy is relatively simple, the debate about it is anything but. Virtually all the big players in the biodiesel industry in the United States and Europe warn about possible engine damage and complain that straight-veg users are breaking the law by not paying fuel taxes and giving the biodiesel industry a bad name when problems do occur. There is some justification for this view, since most of the general public is still not educated enough about various fuels to understand the finer points of these different strategies. On the other side, the straight-veg camp (which also includes many backyard biodiesel producers in the B100 Community) tends to have a conspiratorial mind-set about the biodiesel industry's intentions and complains that the industry is spreading false information about straight veg. Unfortunately, the print and broadcast media cause additional confusion when articles or news reports confuse biodiesel and straight vegetable oil in their reporting. This happens frequently.

The straight-veg strategy appeals mainly to people who are mechanically inclined and who can remember to switch between fuel tanks at the appropriate moments (this is definitely *not* a good strategy for the absent-minded). If you enjoy spending your weekends under the hood of your vehicle—or under the cover of a dumpster scrounging for smelly, greasy, used cooking oil—this may be the strategy for you. However, many people who experiment with SVO, especially those who use their vehicles primarily for around-town driving, eventually give it up because the system never warms up enough to be of much use, and they find that it's simply too much trouble. In addition, there may be some potential long-term engine problems, especially if the straight veg is used carelessly.[48]

SVO Heat

In addition to its use in vehicles, it is also possible to burn SVO/WVO in a conventional home-heating furnace or boiler, but equipment modifications are definitely necessary. Since WVO is too thick to flow through and combust properly in a conventional heating system, it must first be pre-heated and filtered. Straight vegetable oil and waste vegetable oil are being successfully burned in adapted residential oil-fired furnace burners such as Arco, Beckett, Carlin, Ducane, Esso, International, Riello, Slant Finn, and Wayne. However—and this is important—making alternations to your home heating system, especially if you have no idea what you are doing, can be a potentially fatal mistake, and I do not personally recommend it. Be sure to educate yourself completely before you even *think* about pro-ceeding. Since the burning of waste vegetable oil in these burners is not UL-approved, you also need to be aware that you may be breaking a variety of laws and codes if you get involved in this strategy. Nevertheless, for those who want to learn more about this approach, the Yahoo altfuelfurnace forum (http://groups.yahoo.com/group/altfuelfurnace/) is one of the best online meeting places to exchange information and ideas about adapting residential oil burners that use conventional number 2 home-heating oil to these alternative vegetable oil fuels. But since there is a limited supply of used cooking oil, straight veg—whether it's used to power your vehicle or heat your home—is always going to be a somewhat fringe option. Nevertheless, the lure of essentially free fuel is hard for some people to resist.

Laughing Stock Farm is a four-season, certified organic farm located in Freeport, Maine, that serves the restaurant, retail, and consumer markets from Portland to Brunswick. Four-season farming in 17,000 square feet of greenhouses allows farm owners Lisa and Ralph Turner to provide fresh, local produce to restaurants and retailers all winter, while their eight out-side acres rest and rejuvenate for the next growing season.

In recent years, Ralph Turner conducted extensive research on the energy options he had available for his greenhouses on the southern coast of Maine. After looking at wood (no affordable woodlot immediately available in his location) and homemade biodiesel (too much trouble and waste in his view), he concluded that the direct burning of WVO (he prefers to call it used cooking oil) in commercially available waste-oil burners was his best option. He then went about developing a project to test his findings, with support from a development grant from the Maine Department of Agriculture.[49]

The key elements that made this strategy possible was his discovery of sev-eral companies that manufacture commercial-sized waste-oil burners for automotive waste oils that were also certified to burn vegetable oil. Most commercial waste-oil burners will not accept WVO because of its glycerin content, which is the same reason it cannot be burned directly in diesel engines. WVO also has a high flash point of about 400°F (205°C), making it extremely difficult to burn completely (the higher the flash point, the more difficult the combustion process). Biodiesel (by comparison) has a flash point of about 260°F (127°C), while petrodiesel's is only 125°F (52°C). Most waste-oil burners use compressed air to assist in atomizing (fine spraying) the waste oil for better combustion, and a small heater block to make sure the waste oil stays liquid during the flow and atomization. But the system the Turners are now using includes a ceramic inner chamber and target to provide a surface at approximately the flame temperature, thus ensuring complete combustion. "After one failed attempt with another system, the Clean Burn system we are now using has become our preferred option because it actually works," Turner says.[50] There were some minor initial problems with the new system, but the Turners discovered that if they simply removed the burner nozzle and cleaned it about once a week, the new system worked as advertised.

The WVO is picked up in 55-gallon drums from area restaurants for free and trucked to the farm, where they are eventually brought into the greenhouse using a tractor and forklift, and then warmed up before the oil can be processed. Once it's warm, the WVO is sucked out of the drums through a basket strainer (to remove some of the food bits) by a pump and discharged into a heated settling tank. The oil is heated and then sits overnight in the tank to allow water and sediment to collect at the bottom. "Originally, we had a 100-micron stainless steel screen in the tank to remove the food bits," Turner says. "It worked, but was fairly labor-intensive. We finally figured out that we didn't need the screen at all. Now we just heat the oil and let all the solids drop out without filtering." The next day, the water and sediment is drained out of the bottom of the tank, and the clear oil is then pumped to the boiler fuel storage tank through another 100-micron screen to eliminate the last of the food bits. The sludge from the bottom of the tank is composted; nothing goes to waste. Once the WVO has been cleaned and heated, the system works pretty much like any other heating system. When the thermostat calls for heat in the greenhouse, the burner in the 350,000 Btu boiler turns on and heats the water in the hydronic heating loop. Turner chose a hot-water boiler to assist in the processing of the WVO.

One of the restaurants on the Turner's WVO pick-up route has a supply of chicken, beef, and pork fat from a barbecue operation, and this fat has been mixed in with the rest of the WVO without any problems. Because of the animal fat content in his fuel, Turner now refers to all of it as "triglyceride fuel," a more technically accurate description. The Turners hope to add one or more hot air furnaces to their greenhouses, because waste-oil furnaces (as opposed to boilers) are much less expensive and have higher combustion chamber temperatures that burn the WVO more easily.

Although the WVO heating project has been somewhat labor-intensive, overall it has been both a technical and financial success, saving the farm over $20,000 in fuel costs in two heating seasons. Local restaurants have saved almost as much in disposal costs. The fuel savings payback time was calculated to be just over three years. If only hot air furnaces had been used, the payback time could have been a remarkable one and a half years. The state of Maine's estimated 1.5 million gallons of annual WVO production is dwarfed by the state's 400 million gallons of yearly petroleum heating-oil consumption. Nevertheless, the study at Laughing Stock Farm concluded

The 350,000-Btu boiler in the Laughing Stock Farm greenhouse. Laughing Stock Farm.

that the WVO heating program could be expanded to include more green-houses, dairies, and other agricultural operations that were willing to work out agreements with local restaurants in their immediate areas. Recently, Ralph Turner has been working with Underwriters Laboratories (UL) to certify the use of WVO and animal fats in waste-oil burners.

"I feel that this is an absolutely perfect solution for small-scale agriculture in conjunction with surrounding local communities," Turner says. "It is far more energy efficient than biodiesel, it uses no hazardous materials, requires no biodiesel processing equipment or facilities, and it provides all of the same air-emissions benefits from combustion as biodiesel. We're very pleased with this strategy." For more information, pay a virtual visit to Laughing Stock Farm at www.laughingstockfarm.com.

Virgin Veggie
There is yet another variation on the straight-veg strategy that does not rely on dumpster diving behind the fast food restaurant: virgin, rather than used, vegetable oil. This strategy is mostly applicable to farmers who want to grow small quantities of oilseed crops to power their own vehicles, farm machinery, and heating systems. This is not a "free fuel" strategy, since it costs time, effort, and money to grow the crops. But it is a valid fuel-security strategy for essential services—at least on an interim basis—in an increasingly insecure and volatile energy market. While this approach does require an expensive oilseed press (which could be shared) and retrofitting equipment to burn SVO, it eliminates the biodiesel-making process (and the headache of what to do with the low-grade glyc-erin by-product of small-scale transesterification). This may prove to be a viable strategy for some members of the agricultural community, either individually or in small cooperative ventures, especially if they are not located near a ready supply of used cooking oil. This would certainly bring Rudolf Diesel's original vegetable oil vision full circle, and he would undoubtedly be pleased.

CHAPTER SEVEN
GEOTHERMAL

People have been using geothermal heat for thousands of years, especially in naturally flowing hot springs. As early as ten thousand years ago, Native Americans used hot spring water for warmth, cooking, and medicine. Our distant ancestors undoubtedly enjoyed a hot soak now and then, especially after a hard day of mammoth hunting. The Romans used hot springs to treat a variety of diseases, a practice that continues to this day around the world, and the doomed city of Pompeii even had buildings that were heated geothermally. Since the 1960s, France has been heating up to two hundred thousand homes with geothermal water.

Geothermal means "earth heat," and there are two main types. When most people think of geothermal heat, hot springs and places like Yellowstone National Park immediately come to mind. This is sometimes referred to as high-temperature geothermal heat. But you don't need to live in a geologically active area to tap into ground heat. The second type of geothermal heat, low-temperature, is thermal energy from the sun that is absorbed by the upper layers of the earth's surface. That's because within the top fifteen feet or so of the surface, ground temperatures remain fairly constant year-round at between 45° and 60°F (7° to 16°C) in most locations, and this moderate heat can be tapped by devices known as ground-source heat pumps. We'll look at high-temperature geothermal first.

HIGH-TEMPERATURE GEOTHERMAL

High-temperature geothermal makes use of heat from the interior of the earth to produce clean, more or less sustainable energy. The ultimate source of this energy is believed to be radioactive decay occurring deep within the earth, but relatively recent volcanism also plays a major role.

This naturally occurring heat is used in two main ways, *direct use* and for the *generation of electricity*. The highest temperatures are generally reserved for electric generation, while more moderate temperatures are utilized for direct-heating strategies. Sometimes geothermally heated water or steam finds its way to the surface through cracks and faults in the bedrock and forms hot springs or geysers. But most superheated water or steam collects deep underground in porous geological formations known as geothermal reservoirs. This geothermal energy is normally tapped by wells drilled into these geological formations and used to bring the hot water or steam to the surface.[1] Once it's at the surface, these hot geothermal resources can be used directly for the heating of spas, greenhouses, aquaculture, district-heating systems, industrial processes, and so on, or for the production of electricity. Many of the countries that have abundant geothermal resources are those bordering the Pacific Ocean in the so-called Ring of Fire. These nations include Chile, Peru, Ecuador, Colombia; all the countries of Central America; Mexico, the western United States, and Canada; as well as Russia, China, South Korea, Japan, the Philippines, Indonesia, Australia, and New Zealand. Other countries located along the Great Rift of Africa and in the Eastern Mediterranean also have good geothermal resources that could meet a substantial portion of their energy needs.[2]

Active volcanoes, plate tectonics, and the "Ring of Fire." U.S. Geological Survey.

Direct Use

Direct use of geothermal heat for various purposes around the world amounts to the equivalent of about 16,000 megawatts of electricity generation. More than seventy-two countries have reported direct use of geothermal energy.[3] Iceland is the world leader, with 93 percent of its homes heated with geothermal energy, which represents a savings of over $100 million annually in avoided oil imports.[4] In the United States, direct commercial use of geothermal heat began in 1830 when Asa Thompson charged a dollar for the use of his hot-spring-fed tubs in Hot Springs, Arkansas. In 1847, William Bell Elliot, a member of John C. Fremont's survey party, stumbled upon a steaming valley just north of what is now San Francisco, California. Elliot, who thought that he had found the gates of Hell, called the area The Geysers (a misnomer, since there are no geysers there). In the years that followed, The Geysers was developed with several resort hotels featuring hot spas (and more recently it has become home to the world's largest and most productive geo-electricity installations).

The construction of the Hot Lake Hotel near La Grande, Oregon, in 1864, however, was the first time that the thermal energy from hot springs was used on a large scale in the United States. The first commercial greenhouse use of geothermal energy took place in Boise, Idaho. The operation used a 1,000-foot-deep well, drilled in 1926. Geothermal Food Processors, Inc. opened the first geothermal food-processing (crop-drying) plant in Brady Hot Springs, Nevada, in 1978.[5] Since then, although the industry has had its ups and downs, the use of direct geothermal heating has proliferated in parts of the country that contain these resources, mainly in the western states of the lower forty-eight, as well as Alaska and Hawaii. The overhaul of the Geothermal Steam Act of 1970 contained in the Energy Policy Act of 2005 has removed some of the most significant impediments to the direct use of geothermal resources in the United States, and there has been significant renewed interest in direct use by ranchers, schools, tribes, and others throughout the West.[6] In the case of Hawaii, however, there has been longstanding opposition to the development of geothermal resources by local environmental and other groups, who maintain that this would damage the tropical rain forest and violate local religious beliefs.

District Heat

One of the most widespread direct uses of geothermal heat is for the space heating of buildings. This strategy can either be used for a single building or for large groups of buildings or entire communities (a strategy referred to as *district heating*). The oldest district-heating system that is still in operation was warming the French village of Chaudes-Aigues from geothermal hot springs beginning in the early fourteenth century. Modern district geothermal systems also heat homes in Russia, China, France, Sweden, Hungary, Romania, and Japan. The world's largest district-heating system, however, is located in Reykjavik, Iceland. Since it started using geothermal energy as its main source of heat, the city—once highly polluted—has become one of the cleanest cities in the world.[7] The first geothermal district-heating system in the United States was constructed in 1892 in Boise, Idaho. Within a few years, the spring-fed hot-water system was serving two hundred homes and forty downtown businesses. Today, there are four district-heating systems in Boise that provide heat to over five million square feet of residential, business, and governmental space. There are now at least eighteen geothermal district-heating systems in California, Colorado, Idaho, Nevada, New Mexico, Oregon, and South Dakota, and dozens more around the world.[8] Although these systems require significant investment for well drilling and system installation, they have lower operating and maintenance costs and do not require constant (and increasingly expensive) fuel purchases.

Here is how it works. In a geothermal district-heating system, geothermally heated water is extracted from a well and then pumped through a heat exchanger, where it transfers its heat to city water in a closed loop that is piped to buildings in the district. Then, additional heat exchangers transfer the heat to the individual buildings' heating systems. The used geothermal water (sometimes referred to as "exit brine") from the well is normally circulated back to its source through a return pipe and is injected down a different well into the reservoir to be heated and used again in a continuous loop. In the western United States there are more than four hundred communities in sixteen states with geothermal resources suitable for this use available within a five-mile radius, generally considered to be the practical outer limit for economic feasibility.[9]

Klamath Falls, Oregon

Located in the south-central region of the state bordering northern California, Klamath Falls, Oregon (population 19,000), sits atop a hot-water reservoir heated by the same geological processes that fuel the Northwest's infamous volcanoes. Taking advantage of that resource, Klamath Falls now has one of the largest district-heating systems in the United States. The system was initially constructed in 1981 to serve fourteen government buildings. During the first ten years of operation, the system experienced a number of technical and legal problems, but has operated dependably since 1991. At peak heating demand, the original buildings on the system used only about 20 percent of the system's thermal capacity, and income from heating those buildings was not enough to cover operating expenses. As a result, the city launched an active marketing effort to attract additional customers, and, since 1992, the customer base has increased substantially.[10] Today, the system provides about 6 megawatts to heat over twenty municipal or commercial buildings and churches and to melt snow over nearly 105,000 square feet of sidewalks and bridges. A three-acre commercial greenhouse was recently added to the system.[11]

"Landing those greenhouses was a major success; the more customers we can add to the existing system the better," says Klamath Falls' city manager Jeff Ball. "A number of years ago, we extended the system down to the sewer plant, and that's where the greenhouses are located. That extended line to the sewer plant goes through a former mill site along the lake, which is under redevelopment as a planned unit development. The developers have a tremendous interest in using the geothermal resource from that line. There are also a handful of buildings located downtown that are now looking at the geothermal system due to the high cost of natural gas."[12]

The geothermal system relies on two production wells. A lower, cooler (206°F/ 97°C) well is used in mild seasons while the hotter (226°F/108°C) upper well is used in the colder months of the winter. The main geothermal supply line runs about half a mile from the supply wells to a pumping station and heat-exchange facility housed in the county museum, where an injection well located nearby is used to recharge the reservoir. A separate, closed loop runs from the heat exchanger in the museum to the city's geothermal heat customers and back again to the heat exchanger. A new Klamath County Government Center, completed in 1998, includes a new backup boiler for the geothermal heating system, although individual system cus-tomers are supposed to provide their own backup heating systems as well. The city's district-heating system received a number of significant upgrades and improvements during a maintenance shutdown in 2003, including new components and modern system controls, as well as a thorough cleaning of (and additions to) the main heat exchangers. The upgrades, which were partially funded with a National Renewable Energy Laboratory grant, improved the reliability and capacity of the system and reduced electrical operating costs considerably.

The geothermal district-heating system in Klamath Falls has operated almost nonstop since 1993, with only short or partial shutdowns for system connections or maintenance, and both city officials and the businesses and organizations that use it are pleased with the system. However, the city's geothermal system still struggles to pay its own way. "Just covering the manpower costs and generating enough revenue to make improvements has been a big challenge," Ball notes. "It's been hard to make the system pay for itself without outside grants." Nevertheless, there have been other benefits, according to Ball. "It's been an incredible tool for putting this community on the map. We constantly have people coming here from all over the world to look at the system and what's been done here," he says.[13]

In addition to the municipal system, the Oregon Institute of Technology, also located in Klamath Falls, uses a district-heating system of its own to heat nearly 700,000 square feet of space in eleven different buildings. The system saves the institute approximately $300,000 per year on heating bills and offers a significant hedge against future increases in fossil fuel costs. Even local residents heat their homes with private geothermal wells that supply hot water at temperatures generally between 190° and 220°F (88° and 104°C). The cost can run as little as $10 per month for both space heating and domestic hot water. There are about five hundred such wells in Klamath Falls; some are shared by several neighbors, while others are for individual family use.[14] Virtually all of these wells use what is known as a *downhole heat exchanger* (DHE). While this requires a specially designed well containing heating loops, the DHE eliminates the problem of what to do with the geothermal fluids, since only heat is taken from the well.

With all of this demand on the local geothermal reservoir there have been problems in the past, which have led to legal wrangling in some cases, according to Ball. "When we started this program twenty-five years ago, we put together a management plan that essentially required individual owners to use downhole heat exchangers, or if they pumped geothermal water, to re-inject the resource back into the aquifer," he says. "As a result, we eventually reversed the decline of the aquifer, to a point where some wells are again flowing naturally that had not flowed in perhaps forty years. I think that's one of our biggest successes."[15]

Canyon Bloomers

Another good example of the direct use of geothermal resources is located near Hagerman, Idaho, the home of Canyon Bloomers (formerly M & L Greenhouses). Located along the Snake River about thirty miles northwest of Twin Falls, Idaho, M & L Greenhouses began operation in 1970 with one greenhouse that used propane and electricity for heating. The greenhouse operation sits atop what is known as the Banbury Hot Springs area. In 1974, a 505-foot-deep well was drilled that supplied relatively low-temperature 107°F (42°C) water, and the greenhouse was converted to geothermal. A second, 1,000-foot-deep well was subsequently drilled that supplies 130°F (54°C) water. Both wells have an artesian flow, although the shallower well occasionally requires a booster pump in order to maintain system pressure. Canyon Bloomers is a contract grower that supplies two thousand varieties

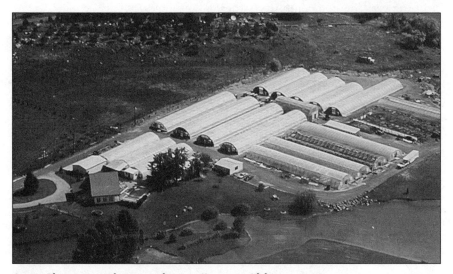

Canyon Bloomers greenhouse complex near Hagerman, Idaho. Geo-Heat Center.

of annual spring plants to large retailers. The growing season starts about mid-December and finishes in late June. In addition to Canyon Bloomers, there are several additional greenhouse operations; a catfish, tilapia, and alligator farm; hot springs spa resorts; and residential heating within about three miles in either direction along the river.[16]

Currently, there are twenty greenhouses of 5,000 square feet each, with a combined area of about 2.3 acres. Geothermal hot water at 130°F (54°C) is used in fan hot-coil heating units, then in radiant floors in sixteen of the greenhouses. The hot water from the first sixteen greenhouses is then circulated in the radiant floors of the remaining four greenhouses. Water is also used in radiant floors in the large office and shop, and in a swimming pool. Three greenhouses have tabletop heating using 107°F (42°C) water and the owner's residence uses mostly 107°F water in radiant floors, but the system can be switched to 130°F water if needed. The total installed system capacity is about a 1.9-megawatt equivalence.

Early on, it was discovered that copper piping corroded quickly and galvanized piping tended to scale and plug. But since the operation was small, the conversion to black iron pipe was fairly easy and inexpensive. Sometimes weak acid is run through the pipes to clean them, according to the owner. Except for the initial pipe problems, the system has performed well. The Canyon Bloomers operation has successfully demonstrated the feasibility of utilizing very low-temperature geothermal resources.[17]

Geothermal Electricity Generation

The first geothermally generated electricity in the world was produced at Larderello, Italy, in 1904. The 10-kilowatt dynamo powered five small electric light bulbs. As of 2005, geothermal energy provided over 8,900 megawatts of electricity in twenty-four countries around the world.[18] The countries that currently produce the most electricity from geothermal resources (in descending order) are the United States, the Philippines, Italy, Mexico, Indonesia, Japan, New Zealand, and Iceland, but geothermal energy is also being exploited in many other countries.[19] In the United States, about 2,800 megawatts of geothermal electricity is generated annually, mostly in California and to a lesser extent in Nevada, Hawaii, and Utah (California's geothermal capacity—2,492 megawatts—exceeds that of every country in the world).[20]

Between 1980 and 1990 a substantial amount of geothermal electric capacity was installed in the United States, mainly in California and Nevada. During the 1990s, low oil and natural gas prices along with other factors resulted in a decline in activity and generation output in the geo-electric sector. However, with higher petroleum and natural gas prices in recent years there has been renewed interest in geothermal energy. In 2000, the U.S. Department of Energy initiated its GeoPowering the West program to encourage development of geothermal resources in the western part of the country. An initial group of twenty-one partnerships with industry was funded to develop new technologies. Geothermal-development working groups in the program are active in eight states: Arizona, Idaho, Nevada, New Mexico, Oregon, Texas, Utah, and Washington.[21] New geo-electric projects are under development throughout the region.

A March 2006 survey confirmed that there is a major surge in developing geothermal electrical power projects in the United States. Some forty-five projects are underway that could nearly double U.S. geothermal electrical-power output, according to the Geothermal Energy Association (GEA), a national industry trade group. The nation had a total of 2,828 megawatts of geothermal power capacity online in 2005. In addition to the states just

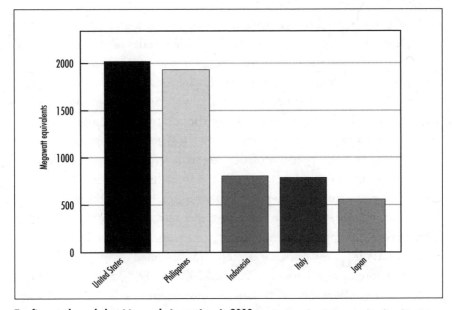

Top five geothermal electricity producing nations in 2003. Based on figures from the International Geothermal Association.

mentioned above, Alaska, California, and Hawaii also have new geothermal power projects, which, when developed, would provide between 1,818 megawatts and 2,095 megawatts of new electric power for the grid. This would be enough electricity to meet the needs of cities the size of Albuquerque, Las Vegas, Sacramento, and Seattle combined, according to the GEA.

"New federal and state initiatives to promote geothermal energy are paying off," says Karl Gawell, GEA's executive director. "State renewable standards coupled with the federal production tax credit are creating a renaissance in U.S. geothermal power production." According to Gawell, the most significant catalyst behind this new industry activity was the passage of the Energy Policy Act by Congress (EPAct) in 2005. EPAct made new geothermal plants eligible for the full federal production tax credit, previously available only to wind projects. It also authorized and directed increased funding for research by the Department of Energy, and gave the Bureau of Land Management new legal guidance and funding to address its backlog of geothermal leases and permits, according to GEA.[22] By virtually every measure, geothermal power is hot.

Geo-Electricity 101

Here is how geo-electricity works. In a geothermal electrical-generating plant, hot steam or water from a specially drilled well is used to spin the turbines that generate the electricity. The used water is then returned to the geological reservoir via an injection well, where it is reheated. This strategy not only helps to maintain pressure, it also assists in maintaining the reservoir—yet over time the heat content of the reservoir gradually declines.

There are three types of geothermal electrical generating plants based on the temperature and pressure of the reservoir it uses. The first is a "dry" steam reservoir that produces steam but very little hot water. The steam is used to turn the turbine generator. Dry steam plants were the earliest type built, and while they are the most cost-effective, they are also fairly rare. The second type produces mostly hot water, and is referred to as a "hot water reservoir." This resource is used in a so-called "flash" power plant: here the heated water, between 300° and 700°F (149° to 371°C) is brought to the surface, where some of it "flashes" into steam in a device called a separator. The steam from the separator turns the turbine. Flash plants are fairly common. The third type, a "binary" power plant, is used where the reservoir temperature is lower, 250° to 360°F (121° to 182°C). In this design, the hot water from the well passes through a heat exchanger, where its heat is transferred to a second (binary) liquid such as isopentane that boils at a lower temperature than water. When heated, the binary liquid turns to vapor, which then spins the turbine. The vapor recondenses to a liquid and is reused repeatedly. In this third type of system there are no emissions to the atmosphere; however, binary plants have higher equipment costs than flash plants.[23]

The vast majority of geo-electric plants are 5 megawatts or larger in generating capacity, although smaller plants are feasible, especially in developing nations where modest initial investments may be preferred. There are approximately fifty geothermal power plants in the world with electrical outputs at or less than 5 megawatts. However, in the United States there are only six plants smaller than 5 megawatts and only one less than 1 megawatt.[24] Nevertheless, small, modular, skid-mounted units can be factory-built and shipped almost anywhere in the world, and because of the modular plant design, a prospective owner can start with a single unit and add additional units later. An additional advantage of small plants is that they can be designed to operate automatically. Where low-cost, shallow wells are feasible, and where the hot exit fluid can be used for direct heating applications, these small units can be economically attractive for powering mini-electric grids.[25]

Geothermal electricity generation is clean, with minimal emissions. The plant requires a very small footprint and, unlike solar or wind systems, it runs constantly, providing baseload electricity. Geo-electric plants are highly reliable, operating up to 99 percent of the time in some cases, compared to 60 to 70 percent for coal and nuclear power plants. Geo-electric facilities, regardless of their size, can also be modular, with additional plants added as necessary. And since many geothermal plants in some countries are locally owned, they can keep the money generated circulating within the local economy rather than sending it overseas for imported oil. On the downside, geothermal reservoirs can eventually be depleted (and can take up to several hundred years to recover). Long-term use of a geothermal site for electricity production requires the construction of new plants at new sites while older sites recover. Also, with "dry" and "flash" plants there are modest CO_2, and sometimes minor NO_x and sulfur-bearing gas, emissions; however, there are strategies that can mitigate most of these releases.[26]

Because high-temperature geothermal sites are not available in many locations, this is admittedly not a strategy for everyone. Nevertheless, there is a good deal of research going on that is focused on drilling very deep geothermal wells into "hot dry rocks" (HDR) in areas not generally considered to be geothermally active. In this HDR strategy, water is injected into one well and extracted from another to recover the heat. So far these experiments have not proved cost-effective. HDR initiatives are primarily being developed in the United States, Austria, France, Switzerland, Germany, and Australia. Although geothermal sources can provide heat and power for many decades, most are eventually depleted, so they are not easily renewable like wind farms or solar installations. Nevertheless, if carefully managed, these sources can provide steady, dependable, clean heat and electricity for a very long time.

The Geysers

The Geysers, located in northwest California near Middletown, is this nation's (and the world's) premier geothermal-electric site. As I mentioned earlier, The Geysers was initially developed for direct use of the hot water near the site in spas and resort hotels. But in 1921, John D. Grant drilled a well there with the intention of generating electricity. His first effort failed, but the following year he was successful at another site across the valley, and the nation's first geothermal electrical power plant began operation. Grant used steam from the first well to drill a second, and several wells later, the operation was producing 250 kilowatts, enough electricity to light the buildings and streets at the resort. The plant, however, was not competitive with other sources of power, and soon fell into disuse.[27]

In the 1950s, Pacific Gas and Electric began to study The Geysers area and conducted a number of preliminary experiments. In 1960, PG&E opened the country's first large-scale geothermal electricity-generating plant at The Geysers. The first turbine produced 11 megawatts of net power and operated successfully for more than thirty years. Many additional wells and turbines have been added over the years, and in 1987, total Geysers steam production peaked at almost 250 billion pounds, enough for an annual average generation of 1,500 megawatts. Two years later, the last

Geysers power plant was completed, bringing the total installed capacity to 2,043 megawatts from twenty-three plant sites. Comprising thirty square miles along the Sonoma and Lake County border, The Geysers is the largest complex of geothermal power plants in the world.

In 1997, in an effort to recharge the declining steam output from the reservoirs, the Lake County-Southeast Geysers Effluent Pipeline Project began operation. This was the first wastewater-to-electricity project in the world. The twenty-nine-mile underground pipeline now delivers 8 million gallons of treated reclaimed water to The Geysers every day to be recycled into the geothermal reservoirs. A similar project delivering 11 million gallons of treated reclaimed water from Santa Rosa was completed two years later. In 2000, The Geysers celebrated its fortieth anniversary and had a total of 350 steam wells and about eighty miles of pipelines in operation. The Calpine Corporation of San Jose, California, now owns all but two of the twenty-one operating power plants at The Geysers.[28]

The biggest challenge faced by Calpine at The Geysers has been to maintain the steam reservoir. "It's essentially a balancing act between production and recharge," says Kent Robertson, director of public relations at Calpine. Nevertheless the future prospects at the site are good, according

Geothermal electrical generation plant at The Geysers, Middletown, California. Calpine Corporation.

to Robertson. "Increasingly, green power production is becoming a public policy goal and geothermal power provides a predictable and dependable supply of electricity on a 24-hour-a-day basis as opposed to other renewable energy technologies that are intermittent and weather-dependent," he says. "Further, the geothermal power generated by Calpine alone is the equivalent of 60 million barrels of oil per year or 2.4 billion barrels over The Geysers' production life. Given this perspective, there is certainly a role for geothermal power in America's diversified energy portfolio."[29]

Small-Scale Geo-Electricity
The renewed interest in geothermal power over the past few years has generated a resurgence of interest in small-scale geothermal electrical-generation projects as well. One of the main challenges in the small-scale category is the high per-unit cost of these power plants. Up until now, most small-scale geopower plants have been custom-built to order, and consequently have been very expensive. But that situation is about to change dramatically if recent experiments by United Technologies Corporation of Hartford, Connecticut, prove to be successful. UTC has been trying to reduce the cost of small-scale geothermal power-generating units by reverse-engineering mass-produced Carrier chiller components to reduce dramatically the cost of production and to allow for easy, modular construction (a *chiller* is a type of mechanical equipment that produces chilled water to cool air). The company has had a successful 200-kilowatt small power plant that operates on waste-heat applications since 2003. UTC has now adapted that design to run on a moderate-temperature geothermal heat source around 165°F (74°C). This contradicts virtually all available literature on geopower—but it works.

The National Renewable Energy Laboratory in Golden, Colorado, has taken a strong interest in UTC's efforts. "We were looking at ways to change the terms of the cost equation for small geothermal power plants because there is a threshold beyond which people say that it's just too expensive," says Gerald Nix, technology manager of the geothermal and industrial technologies programs at NREL. "We were looking at increases in performance and decreases in cost, and what really attracted me to the United Technologies concept was the fact that they were using off-the-shelf equipment. They are taking existing chiller equipment and running it in reverse with very minor modifications, which means that they are starting with a mass-produced system with the potential to get the cost down to somewhere around $1,000 per kilowatt. That represents a dramatic breakthrough."[30] What's more, since the power plant is essentially the same device as existing production-line chillers, if something goes wrong with the turbine generator module, it can be easily removed for repairs and replaced with another. "It can be serviced by a local Carrier repairman who is already familiar with chillers, so the maintenance infrastructure is already in place," Nix adds. Best of all, this exciting new power plant is being tested in a real-world demonstration project in Chena Hot Springs, Alaska, a very small community that offers a potential model for hundreds, if not thousands, of other communities around the world.

Located sixty miles northeast of Fairbanks, Alaska, Chena Hot Springs first became famous for curing crippled prospectors of their aches and pains in 1905. By 1911, the property boasted a stable, bathhouse, and twelve small cabins for visitors, and Chena was on its way to becoming one of the premier resorts of interior Alaska and a favorite getaway spot for residents of Fairbanks. Today, the 445-acre resort offers an outdoor hot-spring-fed lake, indoor heated pool and hot tubs, lodging, a restaurant, a unique ice museum, spectacular scenery, and a wide range of wintertime and summertime activities.[31] However, it is Chena's semiremote location thirty-three miles off the traditional power grid and abundant natural resources that make it ideally suited to showcase the possibilities of renewable energy, including micro-hydro, wind, photovoltaics, hydrogen, and geothermal power technologies. In addition to the spa, the resort uses its geothermal resource for other direct purposes, including heating three greenhouses and cooling the ice museum through the use of a unique absorption chiller that provides year-round refrigeration.[32]

The resort, with its sixty-five employees, is essentially a small town. "We built and maintain ten miles of road and a 3,500-foot-long runway, a water system, a sewer system, power generation, and a landfill for municipal waste and septage," says Chena Hot Springs owner Bernie Karl, a passionate renewable energy advocate. "We have our own microwave phone system, we also became an electric utility recently, and we're soon to become a wireless phone utility to provide wireless service within fifty miles of Chena Hot Springs. In addition, we will be making hydrogen on-site and hope to have all of our vehicles running on hydrogen by the end of 2006." Karl is committed to powering the resort entirely with renewable energy. "By the end of the year it's our goal to be as close to being totally self-sufficient as a community can be," he says.[33]

To help power all of this renewably, Chena Hot Springs and the U.S. Department of Energy jointly funded an exploration project that was designed to locate and characterize the geothermal resource underlying Chena Hot Springs. The resource was tested by drilling a 4,000-foot-deep well, sited to intersect the geothermal reservoir. The exploration was successful, and, as a result, Chena entered into a partnership with United Technologies Corporation to demonstrate their moderate-temperature geothermal power-plant technology.[34] UTC has tested the plant for more than 1,000 continuous hours in the laboratory, and Chena Hot Springs is the first demonstration project. If the demonstration is successful, the small UTC power plants could be available commercially by the end of 2006.[35]

The Chena Hot Springs project calls for the installation of two binary power plants, each generating 200 kilowatts to power the resort, bringing the initial on-site generating capacity to 400 kilowatts. But there's room for more. The new 60-by-150-foot generation building has reinforced concrete floor space for up to 20 megawatts of power production. The first power plant arrived in June and an official commissioning ceremony took place in August 2006. Up to now, the resort has generated its own electricity with diesel-powered generator sets (that gulp 430 gallons of diesel fuel a day) connected to a large bank of batteries. The cost of electricity generation for the resort is expected to fall from around thirty cents to only five cents per kilowatt-hour. There should be sufficient extra generating capacity with the new geopower system to allow for the production of hydrogen for the resort's vehicles. Chena Hot Springs is the lowest-temperature geothermal resource used for power generation in the world, and it is hoped that this will be the first step in developing Alaska's considerable geothermal resources.[36]

"Geothermal is going to make a really significant contribution to the state of Alaska," Karl predicts. "We are the first geothermal project running in the state, and we are the first geothermal power project in the world making electricity from 165° water. I believe that, with UTC's help, we will make the most significant contribution to power generation in the twenty-

United Technology's small, 200-kilowatt, geothermal power plant, Chena Hot Springs, Alaska.
Chena Hot Springs.

first century, and I am very pleased to be a part of it."[37] That may be a bit of an overstatement. But, then again, maybe not. One of the biggest expenses in a geopower project is the well. There are literally thousands of older oil and gas wells in the United States that produce mostly hot water these days rather than oil or gas. The water temperature from these wells averages around 265°F (130°C), according to Karl. Think about that. The potential is enormous.

Small geopower projects have also been successfully installed and operated in other countries around the world, and this was one of the areas of strongest activity during an otherwise slow period in geothermal activity in the United States during the 1990s. A 1999 report by the Geothermal Energy Association indicated that as many as thirty-nine countries have geothermal resources capable of supplying all of their electrical needs. A small 300-kilowatt geothermal power plant located in a rural agricultural setting near Chang Mai in north-central Thailand has been operating successfully since its installation in 1989. This plant, owned by the Electrical Generating Authority of Thailand, also uses a battery storage system to reduce on-peak electrical demand. The Chang Mai plant is the only geothermal power system in the nation.

Another unique small geothermal facility is located about two hundred kilometers north of Lhasa, Tibet. This plant, owned by the Nagqu Power

Bureau, has the distinction of being the world's highest, at 4,500 meters (14,763 feet) above sea level, and also the only completely stand-alone, off-grid geothermal power project (except for the new Chena Hot Springs project). Installed in 1993, it consists of two production wells and one 1.3-megawatt, air-cooled, binary modular power plant. The geothermal system replaced several small diesel generators. The above examples, and other systems located in various countries, have demonstrated that small-scale geothermal is both technically and economically feasible.[38] There is no question that, given the right geothermal resources, many communities around the world could produce their own electricity and hot water for commercial purposes on a relatively small scale.

LOW-TEMPERATURE GEOTHERMAL

Now we come to the second major category of geothermal heat, *low-temperature*. Unlike high-temperature geothermal, this resource is available virtually everywhere. Relatively low temperature latent ground heat can be recovered by a heat pump. Geothermal exchange (often referred to as *geoexchange*) heat pumps can tap into this resource and heat or cool a building through a system of pipes and ducts. Unlike most other heating devices, heat pumps are not based on combustion. Instead, heat pumps move heat from one location to another. In the winter, heat is extracted from fluid in pipes buried in the ground (or sometimes from pond or well water) and then increased with the help of a compressor. The heat is then distributed through the home or building, usually by a series of ducts. In the summer, the process can be reversed, and heat is removed from the home. Geothermal exchange is the most energy efficient, environmentally clean, and cost-effective space-conditioning system available, according to the Environmental Protection Agency. Best of all, heat pumps can be installed in virtually any kind of building almost anywhere, although there may be site-specific limitations in some cases.[39]

The concept behind geothermal heating is simple: the earth is a huge heat-storage device. For millions of years, the earth has been absorbing and storing solar energy in the air, water, and ground. This stored energy offers enormous potential to meet a major portion of our energy needs. The trick up to now, however, has been to figure out practical ways of harvesting and using that energy.

Heat pumps have been around since the early 1900s—refrigerators and air-conditioners are types of heat pumps. Heat pumps for space heating were first developed in the late 1940s, but were not successfully commercialized until the 1970s. Although they have been available for over thirty years, heat pumps are still not understood or appreciated by many people, especially those who live in colder climate zones. The term "heat pump" is used to describe a variety of heating systems, and this can be confusing. There are two main types: air-source heat pumps and ground-source heat pumps. Air-source are not strictly considered to be geothermal, so we will only focus on ground-source heat pumps (air-source heat pumps, however, are an excellent way to heat and cool your home, especially if you live in a temperate climate).

There are three sources of heat for ground-source systems: the ground, well water, and surface water. The ground and well water may not sound like logical places to look for heat, but they actually are an excellent source because their temperature remains relatively constant year-round. The temperature of the ground is about 45° to 60°F (7° to 16°C) just a few feet below the surface. The temperature of surface water in rivers, lakes, or ponds varies more than well water, but surface water is a viable heat source in certain situations. The temperature of underground water ranges between 39° to 55°F (4° to 12°C). Geoexchange technology can tap into the latent heat in these sources and can be used worldwide in almost any climate and location as long as electricity is available.

Geoexchange systems consume 25 to 50 percent less energy than traditional oil and natural gas systems, and up to 70 percent less than electric heat and air conditioning. Geoexchange systems can be used in single-family residences, apartments, condominiums, or commercial buildings of virtually any size. The world's largest geoexchange system is located in the Galt House East Hotel and Waterfront Office Building complex in Louisville, Kentucky. The 6,000-ton capacity system heats and cools over 750,000 square feet that includes 600 hotel rooms, 100 apartments, and 120,000 square feet of public areas. With this kind of design flexibility, it should come as no surprise that geothermal heat pumps have attracted a lot of attention in recent years, and their use has been growing at an annual rate of at least 15 percent, with over 600,000 units installed in the United States. New installations have been occurring at a rate of 50,000 to 60,000 per year, the largest growth in the world for geothermal heat pumps.[40] The U.S. Energy Policy Act of 2005 has generated even greater interest with its various incentives for the installation of geoexchange systems, including up to a $300 tax credit for new systems in single-family residences.

How a Heat Pump Works

As any high school physics teacher would tell you, heat naturally flows from a warmer area (or substance) to a cooler area (or substance). Heat pumps, however, can force heat to flow in the *opposite* direction with the aid of a small amount of electricity and a compressor. It's a little like pumping water uphill. A heat pump transfers or "pumps" heat from a natural source, such as the ground or well water, to the interior of a home or business during the winter. While most people don't think of these sources as being especially "hot," they do, nevertheless, contain useful heat that is continuously replenished by the sun.

Heat pumps can also provide domestic hot water, humidity control, air filtration, and, best of all, cooling during the summer. When the heat-pump process is reversed in the summer, it moves warm air out of your house and into the ground where it is dissipated. The ability of a heat pump to double as a cooling device is a real advantage, especially in warmer climates, and eliminates the need for a separate air-conditioning system. One of the most important things to remember about a heat pump is that it does its work at such a high efficiency (up to 400 percent) because it is primarily *moving* heat from one place to another rather than *creating* heat through some form of combustion. This unique design is what sets all heat pumps apart from most other heating strategies.

Here is how ground-source heat pumps operate. Simply stated, the heat-pump cycle begins as cold liquid refrigerant passes through a heat exchanger and absorbs heat from a relatively low-temperature heat source (water or ground). The refrigerant evaporates as the heat is absorbed, becoming a gas. This refrigerant gas then passes through a compressor, where it is pressurized, raising its temperature to over 160°F (71°C). The heated gas then circulates through another heat exchanger, where heat is removed from the gas and transferred to water or air that is circulated into your home or other space via a hot-air duct system or a hydronic distribution system. The temperature of the heated air or water is about 100°F (38°C). As it loses heat, the refrigerant gas changes back to a liquid. The liquid is cooled as it passes through an expansion device and the heat-pump cycle is complete, ready to begin again. The key component inside the heat pump that controls whether the system heats or cools is called a four-way valve. The valve controls the direction of flow within the system. In the heating mode, refrigerant moves one way; in cooling mode it flows in the opposite direction. Geoexchange heat pumps can also filter your home's air and provide humidity control.

The most common heat pump sizing measurement is the "ton," which has nothing to do with how heavy the pump is. This archaic term is a holdover from the days when refrigeration units were used mainly to produce ice for old-fashioned iceboxes that kept food cool. A "one-ton" unit could make a ton of ice in a day. In today's terms, a one-ton heat pump can generate 12,000 Btu of cooling per hour at an outdoor temperature of 95°F (35°C), or 12,000-Btu heat output at 47°F (8°C).

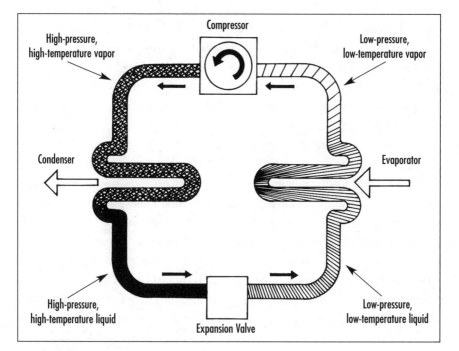

Basic heat pump operation. David Smith.

Environmental Considerations

Because heat pumps supply lower-temperature air for distribution than a conventional fossil-fueled furnace, heat pumps tend to run for longer periods of time, but provide more even heat. For people who are not used to this operational trait, it may appear that the heat pump is "always running." The lower-than-usual temperature of the air that blows out of the heat registers can also be disconcerting to the uninitiated. Yet, a properly designed and installed heat pump will deliver steady heating with less energy consumption than the fossil-fueled competition. This translates into substantial operating-cost savings, especially in areas where fossil fuel prices are high (everywhere these days) and electricity prices are relatively low. Savings are even greater with variable-speed heat pumps. These units adjust their output to match the actual heating or cooling needs of your home, reducing excessive cycling and wear on the system, especially during mild weather.

Because heat pumps consume less energy overall than conventional home-heating systems, heat pump use helps to reduce greenhouse gases such as carbon dioxide, sulfur dioxide, and nitrogen oxides. However, the overall environmental impact of heat pumps depends on how the electricity they consume is generated. (Electricity from renewable sources is obviously preferable to electricity generated by fossil-fueled facilities, especially coal-burning power plants.)

Since there is no combustion involved, heat pumps tend to be very safe, with virtually no risk of fire or combustion gases escaping into your home's living space. Because they are equipped with a wide range of electrical devices and controls, heat pumps require electricity to function, making them susceptible to power failures (most other modern mechanical heating systems suffer from the same problem). Heat pumps are particularly suited for use in new construction, but can also be used in many retrofit situations. Admittedly, heat pumps use a lot of technology that could be largely avoided in a home with an intelligent passive solar design. However, passive solar does not work well on cold, cloudy winter days in northern climates. Heat pumps, on the other hand, do not suffer from this shortcoming, and are an intriguing combination of technology and renewable resources.

Heat-Pump Systems

On the outside, heat pumps are unexciting; they're rectangular metal boxes with pipes and wires coming out of them. It's what's inside the box that counts. Not all heat pumps are created equal, and there are so many choices. Unlike an air-source heat pump, where one heat exchanger (and frequently the compressor) is located outside, a ground-source heat pump unit is located entirely inside your home, often in the basement or utility room. This is especially important in northern climates, where extremely cold winter temperatures and snow can have an adverse impact on heat-pump equipment located outdoors. Part of the heating system is located outside, however, and that is the underground piping that contains the liquid heat-exchange medium. The type and configuration of the outside piping system is the main factor that differentiates various ground-source heat-pump designs.

Types of Ground-Source Heat Pumps

Ground-source heat pumps can be open-system or closed-loop. These terms refer to the design of the piping system located outside the home. An *open system* uses the latent heat in a body of water, which is usually a well but

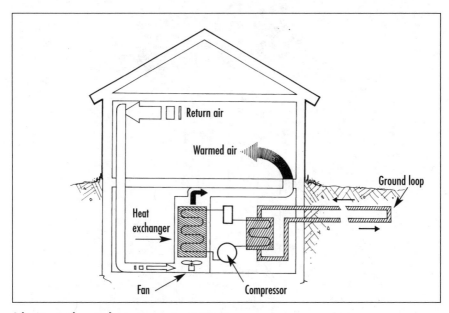

A basic ground-source heat pump system. David Smith.

sometimes a pond or stream, as its heat source. The water is pumped from the well or other source to the primary heat exchanger in the heat pump, where heat is extracted. The water is then discharged into a pond or stream or into a separate (or sometimes the same) well. In a *closed-loop system*, however, heat from the ground is collected by means of a continuous loop of piping buried underground. An environmentally safe antifreeze solution circulating through the piping absorbs heat from the surrounding soil. The antifreeze solution is then drawn into the primary heat exchanger in the heat pump, where heat is extracted. A variation on the closed loop is a *direct-expansion loop*, also called a *DX system*. In a DX system, refrigerant runs directly from the heat pump to the underground piping without passing through a heat exchanger, increasing operating efficiency by 10 to 15 percent.

Closed-loop systems are further subdivided into vertical-loop and horizontal-loop designs. In locations where yard space is at a premium, such as in a suburban subdivision, a vertical loop may be the best design. U-shaped loops of special piping are inserted into six-inch-diameter holes

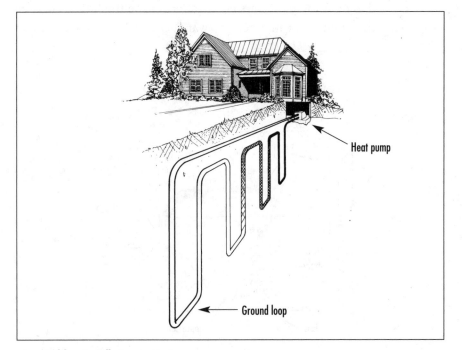

Heat pump

Ground loop

A vertical-loop installation. David Smith.

that have been bored sixty to two hundred feet deep. The length of the loop piping depends on your home's heating and cooling load, soil conditions, climate, and landscaping. In more rural locations, where yard space is not an issue, the horizontal-loop design is more popular. In this strategy, the special plastic pipe is placed in trenches that are from three to six feet deep (the depth depends on the climate zone as well as the number of pipes in each trench). The most common design is to lay two pipes side by side in a trench, but more pipes can be installed if the size of your lot is limited.

"Don't assume that, just because you see a particular geothermal design in a national magazine, it will work in your own state," cautions Harold Rist II, the owner of Smart-Energy of Queensbury, New York. "The horizontal systems don't work when the frost goes down more than about two inches. They show them all over these magazines, but they are talking about the Sunbelt." Rist, one of the most experienced geoexchange designers and installers in the country, has been in the business for over thirty years. In the Northeast, Rist says that a deep, drilled-well system, known as a vertical standing water column, is the most effective strategy. This well can also supply domestic household water. "Of the over six hundred systems we've installed, we've only done six closed loops, and we discovered that the performance with the vertical standing water column was 15 to 20 percent better than with any other system," he says. However, in

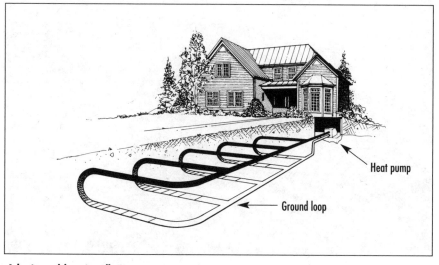

Heat pump

Ground loop

A horizontal-loop installation. David Smith.

other parts of the country—especially where deep, heavy red clays predominate—the vertical standing water column strategy is less effective owing to clay-contaminated water, according to Rist.[41]

In some ground-source systems, a special heat exchanger (sometimes called a *desuperheater*) is used to heat domestic hot water. On some models, the heating of domestic hot water takes place only under certain operating conditions, while on others it occurs on demand. Excess heat is always available during the cooling mode and often available in the heating mode when the heat pump is not working at full capacity. The ability of a ground-source heat pump to provide domestic hot water as an "extra," in addition to space heating and cooling, makes these systems especially attractive. The entire geothermal heating and cooling system produces no smoke, soot, carbon monoxide, hydrocarbons, or thermal pollution of any kind, winter or summer, although the source of the electricity to run the system ultimately determines just how green it really is.

Residential Geoexchange

George Hagerty has used a ground-source heat pump in his home for over twenty years. In fact, he built his 2,400-square-foot Colonial home around one. Hagerty, a mechanical engineer, lives in Queensbury, New York.

Geothermally heated home in Queensbury, New York. George Hagerty.

Although he had been familiar with the heat pump concept as early as the 1960s, Hagerty says it wasn't until he built his new home in 1984 that he was able to incorporate a ground-source heat-pump system into the plans. "I designed the home around the geothermal system," he recalls. "That's because you have to move a lot more air with a heat pump because you get lower-temperature energy at 95° to 104°F (35° to 40°C) versus 140°F (60°C) with an oil- or gas-fired heating system. There are some interior partition walls in the house that actually have 2-by-6-inch studs in order to accommodate the larger ductwork. We were fortunate getting the house set up right from the beginning for any kind of geothermal system in the future."[42]

Hagerty had a contractor install the ductwork and supply the heat pump. But owing to his experience and interest in mechanical engineering, Hagerty did the actual heat-pump installation himself. The original system, however, did not have the four-way valve feature typically found on heat pumps today, and instead used a separate water-to-air heat exchanger for summer cooling—a strategy that left something to be desired. "I was able to get some cooling with that heat exchanger, but nowhere near what I wanted for fast recovery. I would have to anticipate a warm day and turn it on ahead of time," he admits. The heat-pump system uses two drilled wells, one for water supply, and the other for return.

Then, in 1999, the original heat pump began to fail, and instead of repairing it, Hagerty decided to replace it with a new, more modern unit that contained a four-way valve that would allow the system to perform both heating and cooling functions more efficiently than the original system. "I've been running that system ever since, and it's been just great," he says. But why did Hagerty choose a heat pump in the first place? There were several reasons. The fact that he was about to build a new home was a key factor. Air quality was another. "Back in the early 1980s, we weren't that far from the late-1970s increases in oil prices," he says. "Since I had to put some sort of heating system in anyway, I wanted to rely on something besides imported oil. Also, I like the idea of circulating a lot of air in a home because there are things that you can do such as filtering the air with a particulate removal system. That's something that I believe in very strongly. Also, when you have a home with good air circulation you will mitigate radon problems. Air quality definitely was a factor."

Hagerty is especially pleased with the summer performance of his new heat pump. "What's beautiful in the summer is that it will remove so much moisture from the air," he says. "Before I had a gravity drain set up I was draining it by hand, and I was emptying five or six buckets a day." Hagerty does have some advice, based on many years of experience, for others who are considering a ground-source heat pump. "A lot of people don't really understand their electrical and plumbing systems, so they need to have somebody who can service their heat-pump system or teach them how to do it properly," he says. "Things like cleaning the coil, which I do periodically, and changing the air filters, which I do about every two months." Then there is the water system. "We have hard water where we live, so I have water filters for particulate matter that I clean twice a year." Hagerty also suggests that having a backup heating system is a sound strategy, especially in colder northern regions.

Nevertheless, Hagerty is an enthusiastic supporter of ground-source heat-pump systems. "I certainly would recommend a heat pump because it's important to eliminate our dependence on oil and gas for heating. People can get electricity from hydroelectric power plants or from wind or other renewables to power their home-heating system rather than burning fossil fuels directly for heat."[43]

Luther College

Luther College, located in the small northeast Iowa town of Decorah, has two examples of successful, closed-loop, geothermal heat-pump systems. One is located in Baker Village, a student housing complex, and the other is in the college's striking new Center for the Arts. Interest in geothermal systems on campus was initiated by a student environmental group, and eventually resulted in the installation of a 72-ton system in the 33,000-square-foot Baker Village complex. A large group of heat pumps ranging in size from 2.5 tons up to 5 tons located in various parts of the complex are connected via underground plastic piping to 88 vertical closed loops in 150-foot-deep holes bored in the ground. The closed loops are filled with a food-grade, biodegradable antifreeze solution that circulates and acts as the heat transfer medium. The two-story complex was opened in 1999, and provides housing for approximately one hundred students.[44]

The Baker Village system was so successful that the college decided to use

Baker Village student housing complex at Luther College, Decorah, Iowa. Aaron Lurth.

geothermal in its new Center for the Arts as well. Opened in April 2003, the two-level, 60,000-square-foot Center for the Arts is home to the college's art and theatre/dance departments, and also contains a 225-seat theater, classrooms, a computer lab, darkroom, faculty offices, two art galleries, and a café. The 248-ton geothermal system includes 52 two-speed heat-pump units ranging from 1 to 15 tons, connected to 86 wells bored to a depth of 300 feet. This geothermal heating and cooling system was the first to be installed in an academic building in Iowa. Although the system was initially more expensive to install than conventional heating and cooling systems, the college expects to recover the added investment in less than five years on account of lower operating and maintenance expenses.[45]

"The two geothermal buildings have been part of an overall energy audit and conservation project that has been conducted by the college over the past two years," says Jerry Johnson, the college's director of public information. "Those buildings have played a major part in the reduction of the college's energy use by around 12 percent compared to two years ago. Obviously, we're very pleased with the performance of the geothermal systems in those buildings." There have been no major operational problems with either system, and the transition for the college's facilities staff was simple and easy, according to Johnson. "Incorporation of the Center for the Arts into the main campus in conjunction with our central heating system went quite smoothly," he says. "Anyone who has a multiple-facility complex like ours who is wondering how they would incorporate a system like this needs to know that it was not as challenging as they might think."[46]

Founded in 1861 by Norwegian immigrants, Luther is a four-year college affiliated with the Evangelical Lutheran Church in America. The college offers a liberal arts education leading to the bachelor of arts degree in sixty majors and preprofessional programs for its 2,600 students. Additional geothermal systems are likely at Luther. "If the college puts up any buildings in the future, I am certain that geothermal will be part of the design and operation of those buildings," Johnson continues. "The college is currently conducting a capital campaign to build a new science building, and although the plans are not complete, I am sure geothermal will be part of that project too. We've been very happy with these systems."[47]

Munithermal

There is a rather unusual variation on the typical municipal geothermal energy system that's a little hard to classify. This strategy, sometimes referred to as *munithermal*, uses a municipal water system as its water-to-water heat exchange medium. There do not appear to be many of these systems, at least in the United States, but they are intriguing nevertheless because they make dual use of an existing infrastructure. The systems draw potable water from a municipal water line into a heat exchanger in the building, where its latent heat is extracted and then the slightly cooled water is returned to the municipal system. The heat exchanger warms the air for space heating in the building. In the summer, the process is reversed, and the warm building air is drawn through the heat exchanger, warms the municipal water and cools the air that is returned to the building. It's a simple but elegant strategy, at least in theory.

A number of cities in South Dakota—including Pierre, Deadwood, and Belle Fourche—installed these systems in the early 1990s. Some of these installations worked well initially, and several even won state and national awards for their energy efficiency. However, over time a number of problems developed, including corrosion of copper fittings due to high concentrations of carbon dioxide, and lower than expected water temperatures during the winter months, which compromised system performance. As a result, some of the systems have been replaced with more conventional heating and cooling strategies. Nevertheless, in Pierre, the city hall still uses its munithermal system for cooling during the summer, but only during the fall and late winter for heating, according to Todd Chambers, Pierre's utilities director. "Part of the problem was that the temperature of our water system in the winter approaches 50°F, and that seemed to be the break point where the system would not operate properly because the water was too cold."[48]

Consequently, there are some limitations and other issues that need to be kept in mind with these types of systems, according to Phil Nichols of Rapid City, South Dakota, who was involved in the original installation of the projects. "When you are designing a system like this, you need to be sure that you can extract enough heat from the water in the winter and dispose of the heat in the summer," he explains. "It's a little hard to predict the flow of an underground municipal water system. In the summertime, if you don't dissipate the heat, you could get concentrations of hot water or even run out of a place to put the heat, so it's important to analyze the movement of the water. In the projects we worked with, we were certain that the heat was being dissipated. It's not really that much heat, but it's like throwing a beer can out of your back door; if you keep doing it, eventually you have a big pile of cans unless you have some plan to dispose of them."

Another potential problem with this strategy is carbon dioxide content in the water, which will eventually destroy copper pipes and fittings. "CO_2 loves to eat copper, and most geothermal heat pumps have copper components, so make sure your water does not contain CO_2," Nichols says. "If you take a water sample and send it in to a testing laboratory, the CO_2 won't show up. The water has to be sampled and tested on-site for CO_2." The final, and perhaps most important, caveat for munithermal systems is water temperature, according to Nichols. "You have to know exactly what your water temperature is ahead of time. If you live in the northern part of the country, you definitely need to know what it is in the winter, because, if it runs too cold, you many not be able to operate the system."[49] In addition, heightened concerns about security of municipal water systems have added another dimension to the equation, and all but rule out nonmunicipal, private ventures. Despite these limitations, there is undoubtedly some munithermal potential in other locations.

Some of that potential has been tapped in Toronto, Canada (home of the WindShare project described in chapter 3), where the city is using a variation on the munithermal idea—lake-source cooling—to keep many of its buildings comfortable in the summer. While this strategy is not strictly "geothermal," since no heating is involved, it nevertheless deserves mention since it is a creative use of a local thermal resource to dramatically cut cooling costs and CO_2 emissions while simultaneously providing drinking water for the city. The idea is to use the extremely cold 39°F (4°C) water from the bottom of an adjacent body of water (Lake Ontario in this case) to provide the cooling medium. The first city to make use of this strategy was Stockholm, Sweden, in 1995, but the capacity of that system is less than half of Toronto's. The concept was transplanted to North America, where it has been used on Cornell University's campus in Ithaca, New York, beginning in July 2000, and now in Toronto beginning in August 2004.

Here is how the Toronto system works. The icy-cold water from 272 feet (83 meters) below the surface of Lake Ontario is drawn through 3.5-mile-long intake pipes by Enwave District Energy to the city's John Street Pumping Station, where the cold temperatures from the lake water are transferred to Enwave's closed chilled water supply loop by heat exchangers. That separate supply loop then distributes the coldness to participating office towers and other buildings in Toronto's downtown financial district. The water drawn from the lake, now at a more usable temperature, continues on its regular route through the pumping station for normal distribution into the city's potable water supply. Enwave uses only the coldness from the lake water, not the actual water, to provide its alternative to conventional air-conditioning.

The Enwave District Energy system has the ability to cool over 20 million square feet at full capacity. Over thirty-five buildings have already signed on to the project, including high-profile structures such as the Air Canada Centre, Metro Toronto Convention Centre, Royal Bank Plaza, TD Centre, Steam Whistle Brewing, and Hudson Bay Company's million-square-foot retail outlet and thirty-two-story main office. Hudson's Bay expects to save $416,000 a year on energy with the new cooling system. Provincial government buildings at Queen's Park are slated for connection to the system in 2006. Enwave is a private corporation co-owned by the city of Toronto and the Ontario Municipal Employees Retirement System.[50]

"This is truly the energy of the future—available today," says Enwave President Dennis Fotinos. "It offers so many benefits to businesses, developers, public-sector buildings, the environment, and the city of Toronto. It's clean, renewable, reliable energy. Compared to traditional air-conditioning, Deep Lake Water Cooling reduces electricity use by 75 percent, and will eliminate 40,000 tons of carbon dioxide, the equivalent of taking 8,000 cars off of the streets of Toronto."[51]

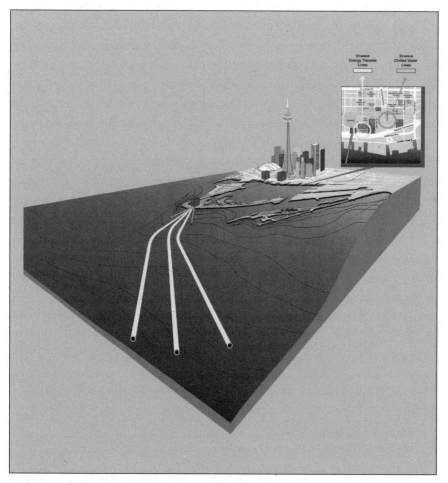

Intake pipes diagram for the Enwave District Energy System in Toronto, Canada. Enwave Energy Corporation.

FINAL THOUGHTS

Although the current use of geothermal resources is substantial, it pales in comparison to the potential. Part of the problem is simply a lack of public awareness. That's beginning to change, as more and more people realize that there are many different types of geothermal resources that individuals and communities can use for heating, cooling, or generation of electricity. And there are many different ways that those resources—both high-temperature and low-temperature—can be utilized. One additional strategy with cooperative possibilities is a ground-source design called *unitherm*, in which a common well (or wells) can be used to supply a group of homes, or perhaps a neighborhood, with geothermal energy. The concept is similar to a high-temperature district-heating system (such as Klamath Falls, Oregon), only in this case it uses a low-temperature resource and also offers cooling in the summer. In this strategy, a common distribution circulator would be used for the whole neighborhood, supplying heating or cooling to the individual houses on the system via a main supply line before returning the geothermal fluid to the original source. Another strategy for a ground-source system, regardless of its size, would be to combine it with a wind turbine or photovoltaic array to create a self-powered hybrid system. There are so many possibilities. Of course, that has been true for virtually all of the renewable energy strategies that we have looked at throughout this book. Now it's time to put all the pieces of the puzzle together—and for us to step back and take a look at the bigger picture and how it relates to community-supported energy.

CHAPTER EIGHT

THE COMMUNITY SOLUTION

It was near the end of September 2005, and as I watched the scenery beside Lake Champlain unfold, I noticed that the foliage outside was late in taking on its bright autumn hues again this year. I was aboard the southbound *Adirondack*, wending its way along the western lakeshore on the way to Schenectady, New York, where I was to change trains for the Chicago-bound *Lake Shore Limited*. As the late afternoon shadows lengthened, I settled back in my seat and pondered the implications of the climate changes that have begun to have such a pronounced—and noticeable—effect on the seasons in northern New England. These were disquieting thoughts.

I was on my way to attend the Second U.S. Conference on Peak Oil and Community Solutions in Yellow Springs, Ohio. My overnight journey to Toledo, Ohio (as close as I could get to Yellow Springs via train), was an adventure in itself, but I made it to the conference as planned in a rental car for the last 160 miles of the trip from Toledo. My decision to travel via Amtrak was partly motivated by my desire to demonstrate my personal commitment to energy conservation, and partly because I simply enjoy riding on trains. During the course of the conference I learned that I was one of only two people there who had traveled via Amtrak to attend. We do love our cars.

The conference, held at Antioch College, was sponsored by Yellow Springs–based Community Service Inc., which through its Community Solution program encourages the resurgence of small local communities in an era of increasingly scarce and expensive oil.

The conference program was designed to help attendees learn how to prepare at the local level for the coming decline in global oil production. There were more than 350 people from thirty-nine states and five countries along with about 100 area residents at the conference, which at the time was the largest event of its kind in the world. The first conference held the previous year had attracted about 200 attendees. As it turned out, this was simply one of the most exciting events I have ever attended, and well worth the long journey.

"Peak oil will undoubtedly be tough," said Pat Murphy, Community Service's executive director, in his opening remarks. "We can make it tougher by trying to hang on to an out-of-date lifestyle or fighting wars for the last drops of oil." The large audience assembled in the packed auditorium listened attentively. "Our theme is the journey home," Murphy continued. "Like the Bible story of the prodigal son who left his community for the lure of the big city, we find ourselves in big trouble. It's time to return to the community, make amends, clean up the mess, and get back on the right path."[1]

Murphy's remarks set the tone for the speakers who followed, including many key leaders in the fast-growing peak-oil awareness movement. Among them were oil-industry expert Jan Lundberg; energy consultant Steve Andrews, cofounder of the Association for the Study of Peak Oil, USA; Richard Heinberg, author of *The Party's Over* and *Power Down,* who gave the keynote address; and many, many others. All of the presentations were excellent, and they covered a wide range of subjects, from how Cuba had survived peak oil to creating intentional communities, and from local food production to the principles of permaculture. But the highlight, for me at least, was the presentation given by John Ikerd, an economist and author of a recent book, *Sustainable Capitalism,* who advocated a new economy based on sustainable energy. Some economists tend to come across as a bit dry, but Ikerd's presentation turned out to be what I can only describe as an electrifying, revivalist sermon, gradually building to an ecstatic crescendo that had the crowd on their feet cheering and applauding. He definitely was preaching to the choir.

During the breaks and especially at mealtimes, there were numerous opportunities to speak with other conference participants and to make connections—and new friends. The diversity of the group was fascinating. There were some slightly shell-shocked individuals who obviously were learning about peak oil for the first time—and who were having some difficulty dealing with it. The more seasoned participants tried to be as supportive as possible; they'd already been through the process and knew how disorienting it can be. There was a large contingent of active community organizers who were attempting to hone their skills or develop new ones. There were even a few rugged individualist/survivalist types talking about stockpiling ammunition and food supplies, who didn't seem to be getting the message about cooperative community effort. But overall, the discussions were friendly, respectful, stimulating, and extremely helpful. The networking that took place was one of the most valuable aspects of the three-day event.

One thing at the conference that struck me, however, was the lack of any discussion about organized, large-scale, community-supported renewable energy initiatives. There was a good deal of talk about community solutions, but no comprehensive community strategy for renewable energy. While there were references to an occasional micro-hydro system or photovoltaic array for a few intentional communities, most of the renewable energy talk was general or focused on systems installed by individuals on their own homes. This seeming paradox is due to Community Solution's strong focus on conservation rather than trying to continue to support our current over-consumptive lifestyle. However, if the primary response to peak oil and global warming is going to be community based, then it seems that a more comprehensive, locally supported, community renewable energy initiative would complement these conservation efforts.

The conference was a fast paced, almost dizzying array of presentations, panel discussions, and ad hoc group meetings—but by Sunday, things finally began to wind down. Pat Murphy ended the conference by saying, "The most important thing is for all of us to 'become the change you wish to see in the world.' If you want a low-energy, caring, community way of living, then become a member of that community—even if you are initially the only member."[2] We were sent home with the clear expectation on the part of the conference organizers that we would all become local advocates for community solutions to peak oil in our own hometowns—a daunting prospect for those with limited experience in such activities. But I suspect that many who attended the conference did not disappoint the organizers, and have risen to the occasion. We'll return to Yellow Springs in a moment.

THE BIG PICTURE

Up to this point, we have looked at the many individual strategies that are available to help us make the transition to a new renewable-energy-based economy. Now it's time to assemble those individual strategies into a more unified picture from which we can begin to identify some larger patterns and draw some final conclusions. As we have seen, each of these strategies has its own set of advantages and disadvantages. As we have also seen, there are no perfect solutions. The trick is to weigh all the potential advantages against the disadvantages to try to determine which strategies are appropriate in different locations and circumstances. And what is appropriate will be different in every community for a variety of reasons, according to the National Renewable Energy Laboratory's Gerald Nix. "Some communities will use geothermal, some will use wind, and some will use biomass," he says. "The answer to our energy challenge is going to be different in different locations. It's going to be a combination of economic factors, local resources, and the inclinations of the people in the area that will be the determining factors." I agree. Let's look at a brief summary of these many strategies, along with their relative advantages and disadvantages, to help us make those decisions.

Solar

As we've seen, solar energy can be used in many different ways to provide heat, hot water, lighting, and electricity. It can be harvested passively by properly designed buildings, or actively by a variety of active systems. Although they are visible, in general, active solar systems are not especially intrusive, particularly if they are carefully integrated into a building's design. Active solar systems are easily scalable, from very small to very large, and are applicable for individual homes or large municipal and commercial installations and everything in between. PV panels quietly produce electricity, do not emit greenhouse gases, and are the cleanest and safest method of generating power at the present time. PV also lends itself to hybrid combinations with wind or hydro systems. Solar energy offers huge potential around the world, but it's not available at night, and is often obscured by clouds in some locations. Sites that lack good solar exposure may not be viable, and the intermittent nature of solar energy is definitely a disadvantage. There is little organized public opposition to most solar strategies, but a lot of ignorance remains about the potential benefits that solar offers.

Wind

Wind power is used primarily to generate electricity. But because wind power is intermittent, it is not well suited to meet constant baseload electricity requirements. Modern wind-turbine technology is highly sophisticated and increasingly competitive with more conventional forms of electrical generation. There is now a wind turbine sized for virtually every use. A good, windy location is best for any wind turbine, and absolutely essential for commercial wind farms. For virtually all wind turbines, regardless of their size, the taller the tower the better. Medium- and larger-sized wind turbines lend themselves well to municipal or cooperative, community ownership, and there are numerous examples around the world. Small wind/PV hybrid systems are especially popular in off-grid locations and are an especially good choice for isolated rural communities. There is organized opposition to large, commercial-scale wind farms (and sometimes even small, individual turbines) in some locations by neighbors, primarily on account of what amounts to aesthetic considerations about visual impact.

Hydro

Hydropower is mainly used for generating electricity. Because most good sites are capable of producing hydroelectricity around the clock, hydro can be used to meet either constant baseload power or peak-power requirements—a major advantage. Hydropower technology is scalable, from very small, individual residential systems to very large, government-sponsored projects. Large-scale hydropower, however, has fallen out of favor in the United States primarily owing to opposition to their associated dams, and the construction of any additional large projects is unlikely. However, there are substantial opportunities for the development of smaller, low-impact sites, which is a good match with the development of local, small-scale, distributed generation. Small-scale hydro lends itself well to municipal- or community-ownership strategies. The use of existing municipal water systems to generate electricity is an underdeveloped resource with considerable potential. A variety of ocean-energy strategies are currently showing a lot of promise, and the fact that they tend to have a low profile (or are completely submerged in some cases) should minimize opposition based on aesthetic visual concerns. Ocean-energy strategies, however, tend to be somewhat intermittent (although tidal energy is highly predictable).

Biomass

Biomass can be used to generate electricity and make solid fuels such as wood chips or pellets for heating or electrical generation (or both in combined heat and power projects). When sustainably harvested biomass is efficiently combusted, it dramatically reduces CO_2 emissions when compared to petroleum, natural gas, and coal. What's more, as the price of fossil fuels continues to escalate, solid biomass fuels have become substantially less expensive than the fossil competition. Biomass strategies are generally scalable, from the individual home system up to the largest industrial-sized facility, making them well suited to cooperative community or municipal systems of almost any size. Landfill gas, farm-based methane, and similar strategies offer additional possibilities that have not yet been fully exploited. The use of syngas or wood gas to power stationary engines (and even vehicles in an emergency) is an interesting option that merits additional attention.

The main limitation for biomass strategies is the availability of feedstock. As long as a sustainable supply of biomass feedstock is available within a reasonable distance, it's a viable strategy. There are some concerns, however, about the potential damage to agricultural and forest lands from the "extensive management" required by the large-scale production of feedstock that is now being promoted. There is also potential for competition between land used for biomass and land used for the production of food. In addition, severe weather conditions caused by global warming could seriously compromise future biomass production. Overall, though, biomass offers enormous opportunity if used wisely.

Liquid Biofuels

A subcategory of biomass, liquid biofuels are going to play an increasingly significant role in transportation, a key sector not well suited to other renewable energy strategies (except in the case of electricity used by plug-in hybrid-electric vehicles for short-distance driving and for

electric-powered trains and trolleys). The primary liquid biofuels—
ethanol and biodiesel—can be produced from a wide range of plant-
based feedstocks, and in the case of biodiesel from animal fats as well
(biodiesel can also be used as a heating fuel). The main advantages of
using biofuels include their ability to be mixed with (or substituted
directly for) fossil fuels, their compatibility with the existing distribution
network, and the lower levels of harmful emissions they produce when
combusted, compared to their fossil fuel equivalents.

Ethanol and biodiesel can be made fairly easily in small quantities using
simple equipment or in large quantities by industrial-scale production facilities. Cellulosic ethanol (based on the whole plant rather than just the
seed) and biodiesel made from algae offer considerable potential for far
greater yields than are possible from more traditional ethanol and biodiesel
production strategies. Biomass-to-liquid technologies also offer quite a lot
of potential. The use of straight vegetable oil (either used cooking oil or
virgin oil) is another strategy for transportation fuel and heating oil that
deserves serious consideration in certain circumstances, especially on the
farm. Although there are substantial regulatory issues involved, both
ethanol and biodiesel offer considerable opportunities for small-scale, local,
cooperative production initiatives, especially at the small family farm or
community level.

There are a considerable number of concerns about liquid biofuels, however, including a low energy ratio for corn-based ethanol as well as a recent
trend in the construction of coal-fired ethanol plants in the United States
(an utterly self-defeating strategy if there ever was one). The explosive
growth of the liquid biofuels sector has also accelerated a movement away
from local or farmer-owned production facilities toward large, corporate-
dominated initiatives with little or no local ownership or control. There also
have been serious concerns raised about the damage caused to farmland by
large-scale agribusiness corn and soybean production, such as soil erosion as
well as pesticide and fertilizer runoffs that pollute waterways. The recent,
growing trend toward "outsourcing" the production of feedstock crops to
Third World countries, resulting in deforestation and other negative environmental consequences, raises additional concerns. The growing reliance
on genetically modified crops as biofuel feedstocks in some countries is
another troubling issue. And, again, potential competition between wealthy

car drivers in developed nations and poor Third World people for feedstock used for both biofuels and food could eventually price the Third World residents out of the market. Last but not least, runaway global warming will almost certainly cause serious disruptions to large-scale agribusiness and may render most optimistic biofuels-production forecasts moot.

Geothermal

There are two main types of geothermal heat: high-temperature and low-temperature. High-temperature resources are used in two main ways: for direct use and for the generation of electricity. The highest temperatures are generally reserved for electric generation, while more moderate temperatures are utilized for direct-heating strategies. The use of direct geothermal heating for district-heating systems in some towns and cities is a popular strategy in areas that have the right geological conditions available. Geothermal electricity generation is clean, with minimal emissions; requires a very small footprint for the plant; and unlike solar or wind systems, runs constantly, providing baseload electricity. Geo-electric plants are also highly reliable, and there have been some encouraging recent developments in the small-scale geothermal arena providing considerable potential for distributed, local generation. Recent high fossil fuel prices and energy legislation has spurred significant new geothermal development in the United States and elsewhere. On the downside, high-temperature geothermal reservoirs are not located everywhere and can eventually be depleted (and may take up to several hundred years to recover), so they are not as immediately renewable as most other renewable strategies.

Low-temperature geothermal, however, is available in most locations and is immediately renewable. Relatively low-temperature ground heat near the earth's surface can be recovered by a heat pump and used to heat buildings in the winter. In the summer, the process can be reversed to provide air-conditioning. Geothermal exchange is the most energy efficient, environmentally clean, and cost-effective space-conditioning system available. Because heat pumps consume less energy overall than conventional home-heating systems, heat-pump use helps to reduce greenhouse gases such as carbon dioxide, sulfur dioxide, and nitrogen oxides. However, the overall environmental impact of heat pumps depends on how the electricity they consume is generated. Geoexchange systems are scalable, from single-family homes to large municipal or commercial installations, and could be used in cooperative situations as well. Variations on the geoexchange strategy such as munithermal for heating and cooling with the use of municipal water systems may offer some additional opportunities in some situations. Because they tend to be virtually invisible and have no major negative environmental impacts, there is no obvious opposition to geoexchange heating and cooling strategies.

COMMUNITY-SUPPORTED ENERGY

There is, however, one major potential problem with all of these renewable energy strategies that is often overlooked by their supporters. While they offer a lot of promise, without strong community support and local ownership, these strategies can simply end up substituting one form of corporate domination for another. This is not much of an improvement, and is at least one reason why some communities oppose large-project proposals. In many cases, community members feel that the project is being imposed upon them by outsiders, and that the local disadvantages outweigh the potential advantages. This may not necessarily be true, but it demonstrates why a direct connection between these projects and the local community is so important. This connection provides the key ingredient that transforms what would otherwise be just another large corporate energy initiative into an engine for local economic development and energy security that directly benefits its owners—the members of the community—rather than a group of absentee investors.

Community-supported energy (CSE) is similar to community-supported agriculture (CSA), except that instead of investing in carrots, tomatoes, or chicken, local residents invest in greater energy security and a cleaner environment. Local ownership and control allows the community to create a project that meets its particular needs while addressing its concerns about size, scale, and location. While community ownership does require local residents to take greater responsibility for their own energy use, it also offers greater returns. Community ownership of renewable energy projects keeps far more money—perhaps three or four times more—circulating in the local economy than absentee ownership.

Don't get me wrong, given the choice between a large, corporate-owned coal-fired power plant or a large, corporate-owned wind farm, the obvious choice is the wind farm. And since the magnitude of the energy crisis we face is going to require a massive response, there is no question that large companies are going to have a role to play. But that's no reason to exclude smaller, more dispersed community projects that are far more effective in promoting distributed-generation strategies. All of these responses will be needed if we are even going to come close to meeting the challenge of replacing fossil fuels in the short amount of time we have available. But being a strong supporter of renewables does not imply, or even require, equal support for a large multinational corporate monopoly on these technologies. More and more people are beginning to recognize that it's the predatory, corporate-dominated, global economic system with its mindless pursuit of short-term profits and "shareholder value" that is devouring the planet and its resources—and ultimately us along with it. Shareholder value. At what cost to everyone else? What "value" is there to a planet that is no longer livable?

Local communities, in contrast, generally tend to be better stewards of their immediate environments because they know that if they are going to continue to thrive they need to conserve those local resources, both physical and human. Yes, there are plenty of examples of communities (or individuals within communities) that are not good stewards, but by and large they tend to be located in impoverished Third World nations that for the most part have been (or are being) decimated by the plundering and indifference of large corporate interests. While the global free-market economy has repeatedly demonstrated that it has no soul or compassion, most communities by contrast are blessed with both. If I had to choose between relying on my community, or some large, faceless, out-of-state corporation for my survival, I'll put my money on the community. And that is exactly what I am proposing. Literally. Why continue to invest in a system that is literally killing us all? Take your time, energy, money, and resources and invest it in your future, and the future of your children, in your own community. Because, as we have seen throughout this book, there are many opportunities and many good reasons to do so, and community-supported energy provides the vehicle.

GROUPS THAT GET IT

As we have also seen, quite a few people understand the benefits of local energy ownership and have already taken advantage of it. There are numerous organizations and initiatives around the world that are strong advocates for direct community involvement in renewable energy projects, especially the wind-power associations of local landowners and residents (*Bürgerbeteiligung*) in Germany, and the many wind cooperatives in Denmark, Holland, Sweden, and England. In Canada, WindShare in Toronto and the new standard offer contracts in Ontario stand out as shining examples of enlightened thinking. Even in the United States, Windustry in Minnesota, the Community Wind Collaborative in Massachusetts, and Solar Sebastopol in California come to mind. Most of these groups or initiatives are primarily focused on specific renewable strategies, but they may not necessarily be motivated by concerns about peak oil or global warming. This is not necessarily a bad thing, but the twin perils of peak oil and global climate change certainly should provide additional incentive for anyone who is not otherwise convinced by the obvious financial benefits of community-based energy initiatives.

There is, however, another group of organizations that definitely *is* focused on peak oil that are natural allies to the concept of community-supported energy—the so-called peak oil community. This loose coalition of organizations around the world is actively promoting what are generally known as *relocalization* initiatives that are aimed at empowering communities to rebuild their local economies and infrastructure to withstand the oncoming Long Emergency precipitated by peak oil and global climate change. While these groups generally have a broad focus that extends well beyond energy, they provide the logical connection between resurgent, self-reliant communities and the energy needed to power them. But in some cases, the renewable energy component is not as well defined as it should be with these groups. Despite the fact that the peak oil community has accomplished a lot in a very short time, it could accomplish even more if it added a more focused, community-supported energy initiative to its existing range of strategies. It's an obvious combination. Some already have done so; others still need to.

The Community Solution

In the absence of any meaningful national leadership, the focus on a local community response to peak oil and global warming is at the very heart of this rapidly growing international relocalization movement. And The Community Solution in Yellow Springs, Ohio, has unquestionably been playing a pivotal role in providing an inspiring model for people to follow. Founded in 1940, the group has long advocated for small, local communities as the most fulfilling and healthy way to live. The Community Solution has more recently been studying peak oil and the coming changes associated with its arrival. The program advocates a transition to small, sustainable communities in response to resource depletion and environmental and social degradation. The Community Solution is focused on public education about peak oil through its Web site (www.communitysolution.org), its quarterly newsletter *New Solutions*, and its annual conference. The group has also produced the documentary *The Power of Community: How Cuba Survived Peak Oil*, developed peak oil workshops for community leaders, and is developing a model post-oil neighborhood community called "Agraria" and an energy information system that helps people to reduce their energy use as efficiently and immediately as possible.

The individual responsible for getting the word out to the public on these many initiatives is Megan Quinn. A graduate of Miami University of Ohio in diplomacy and foreign affairs, Quinn is outreach director for Community Service, Inc., and an articulate spokesperson for the movement. A dramatic increase in the group's activities has been keeping her very busy in the past few years. "Since we started over four years ago, the growth of interest in our programs really has been exponential," she says. "Conference attendance is up, our Web traffic has been increasing, and our membership keeps growing. I think people are becoming more aware, partly because of the increase in gas prices, and partly due to increased coverage in the media about peak oil and global climate change. There is also a strong grassroots movement happening, with people showing *The End of Suburbia* and our Cuba film, giving presentations, and organizing community groups. All of this has been progressing very quickly and I think it will continue to do so."[3]

The growing public awareness of peak oil and global warming, and the direct and indirect connections between them, is one of the key factors driving the movement, according to Quinn. "I think that people are starting to link peak oil and global climate change more and more," she says. "They are beginning to see that you can't effectively respond to one without thinking about and dealing with the other. However, I believe that peak oil is seen as the more immediate issue. Most people still see climate change as being way out there sometime in the future, while peak oil hits home with more immediate economic implications."

Although the early stages of peak oil already seem to be upon us with high gasoline prices in the United States, what if the peak oil community is wrong about its imminent arrival? "It's our view that peak oil isn't really the problem, it's our way of life that's the problem," Quinn responds. "It's not just that our way of life is unsustainable, it's also not increasing our happiness with materialism and its associated high consumption of energy. Considering the deteriorating state of the environment, growing evidence of rapidly increasing climate change, and the threat of resource wars, I think we're going to need to adopt most of these strategies anyway, and peak oil may encourage us to do it more quickly. I think it's important that the movement doesn't get caught up in trying to predict the exact date; it's just a general trend that clearly demonstrates the need to change our society and economy to consume a much lower amount of energy."

One issue that Quinn frequently has to deal with is the negative reaction that many people experience when they first learn about peak oil. "I think that's a natural reaction; in some cases they are in denial," she says. "I think that it can be so overwhelming that it can cause people to become paralyzed. I would recommend that at some point they stop reading the books and Web sites, and watching the movies. Then they need to take a look at their life and analyze how they use energy, and how much they use, and begin to deal with it one thing at a time. They can begin to take small steps and try to consume less energy today then they did yesterday, and less tomorrow than they used today. They can also start to make local connections in their community to work on these problems in a group."

Looking ahead to the future, Quinn is generally optimistic about our prospects, especially if people take advantage of the challenges ahead to create a more sustainable, equitable society that places greater emphasis on community and cooperation. She also recognizes that her generation needs to play a more active role in this process. "As a young person, I believe that the youth of this nation need to begin to take a leadership role now, and I think that we need to realize that we are not going to live the same kind of lives that our parents lived," she says. "We need to come up with a better vision and find better goals. I think the future is resting on the shoulders of my generation."[4]

One of the most effective ways to cut energy use dramatically is to live where you work (or work where you live) and to get most of your food and other essentials from a local source. The Community Solution's proposed post-oil community, "Agraria," to be located in Yellow Springs, puts these ideas into practice. Agraria is designed to be an attractive, low-energy community with clustered, highly energy-efficient homes that will serve as a model for similar development across the country as a response to peak oil and global warming. The organic gardens, low-energy building techniques, and other aspects of the neighborhood-community design will be strong educational tools and even sources of income for some of the neighborhood's residents.

The main idea is for Agraria's residents to live and work within the community to the greatest extent possible. Some residents might work on premises to maintain gardens, grounds, common buildings, and shared equipment, or may run educational programs from the community. Common space includes business space (offices and shops) and business incubator space as well as land for farming and orchards. The plan envisions a major shift to the growing and storage of local food, and the associated skills needed will be encouraged and developed. Provisions in the covenants of the community will ensure that the people who accomplish these tasks are always themselves residents of the neighborhood-community.[5]

A comprehensive, community-based renewable energy strategy is an essential part of the Agraria plan. The community intends to use one-fourth the average current per capita consumption of energy in the United States. This goal is to be realized through carefully designed, smaller (approximately 1,000-square-foot), highly insulated, energy-efficient homes. Kitchens will use highly insulated refrigerators and freezers. Flash water heaters or solar water heaters will replace conventional gas-fired water heaters. Heat pumps may also be used, possibly powered by PV solar cells. A number of other community energy strategies, including wind power, are also being considered.

The steep reduction in energy use envisioned for Agraria is not viewed as either a sacrifice or a burden, but rather as part of an entirely new way of living that offers many positive benefits, according to Quinn. "We want to help develop the case that people shouldn't just reduce their energy use because they have to; it shouldn't be viewed as having only a negative effect on their life," she says. "We want to help people to realize that this could be a positive thing, especially in terms of developing strong communities where interactions and relationships and cooperation are the most important values. We need something better than the present system to dream about, and Agraria is a manifestation of that positive vision."[6]

A fundraising campaign is under way to help move Agraria from the drawing board to reality, and if all goes well, the project could get under way sometime in 2007. For additional information about Agraria, visit www.communitysolution.org/agraria.

Post Carbon Institute

If The Community Solution in Yellow Springs, Ohio, provides an inspiring model for people to follow, then the Post Carbon Institute in Vancouver, British Columbia, supplies the network and additional tools to help peak oil groups transform that model into concrete local action. The Post Carbon Institute (www.postcarbon.org) was established in 2003 as an initiative of the MetaFoundation, an organization created three years earlier, to serve as an umbrella group for new organizations focused on innovative environmental solutions. PCI's mission is to assist societies in their efforts to relocalize communities and adapt to an energy-constrained world. The group believes that production of oil and natural gas will peak soon, that climate change is worsening, and that the current global economic system is unstable and reinforces huge disparities. PCI's response is to promote drastically lower consumption, greater local self-reliance, and more cooperative and inclusive communities.

To accomplish this ambitious agenda, PCI makes use of a number of powerful tools, but especially its Relocalization Network, a rapidly growing consortium of local Post Carbon groups in at least eleven countries around the world. This Internet-based network, with over 115 groups currently participating, is assisting local communities to prepare for a future in which wise use of local resources may very well play the crucial role in determining success or failure. "More and more people understand the implications of the coming crisis, and now they want to know what to do about it," says Celine Rich, PCI's cofounder. "The answer to this crisis is relocalization—strengthening the local community at all levels. And rather than feeling that they are out in the wilderness by themselves, this network helps them look to other groups that are doing the same thing, and to understand what they can do in their own community. Post Carbon Institute is helping to both connect them and give them ideas about what they can do."[7] The network is an increasingly popular strategy, and new Post Carbon groups affiliated with the network continue to pop up all over the world like mushrooms after a summer rain, despite (or perhaps due to) the inattention of most national governments to these pressing issues.

Relocalization is also intended to help reverse the corrosive effects of corporate-dominated globalization, which has devastated local economies and communities around the world and precipitated a global race to the bottom for millions of working people everywhere. To assist it in its work, PCI recently received a $125,000 grant to help establish two new initiatives. The first is a Municipal Action Plan project that will assist municipalities in their efforts to "power down" in preparation for the coming expected reductions in energy supply. When completed, the plan will provide municipalities with concrete ways to reduce their energy requirements by relocalizing the production of goods and services. The second initiative is the Oil Depletion Protocol, which provides a way for nations of the world to cooperatively prepare for, and respond to, the severe consequences of the global decline in oil production.[8] Originally drafted by the Association for the Study of Peak Oil, the protocol would see all oil-importing nations agree to reduce their imports by a specific yearly percentage, while exporting nations would agree to reduce their oil exports. This will not be an easy sell. Fortunately, it's not necessary for all nations to ratify the protocol for it to have a positive effect (the United States is not expected to be first in line to ratify). But even if only one nation signed up, that nation would benefit. However, if a larger number signed on, it would provide a platform for international economic stability and cooperation.[9] PCI is helping to promote the concept to anyone (or any country) that will listen.

"We really should be thinking about greatly cutting our demand and then maybe we'll all have a decent future," says Julian Darley, cofounder of PCI and coauthor of PCI's new book, *Relocalize Now!: Getting Ready for Climate Change and the End of Cheap Oil.* "It's going to be a bit difficult, there is no denying that, but if we work hard and try to think long-term, then we would definitely embrace trying to cut our demand. This admittedly runs counter to what we have been thinking for the last three hundred years in North America. Ever since the European settlers arrived here we've regarded this place as a sort of bottomless barrel of resources, and that idea has really got to be challenged if we are going to get through this success-fully."[10] Darley, who also authored *High Noon for Natural Gas: The New Energy Crisis,* was first alerted to the concept of peak oil while reading Jeremy Rifkin's book *The Hydrogen Economy.* After quite a bit of research, Darley realized that peak oil was the most immediate danger facing the global economy, leading to a profound shift in the focus and work of the MetaFoundation, and its ultimate replacement by the Post Carbon Institute as the main umbrella organization.

Darley offers a few suggestions for people who want to know what they can do about the challenges that lie ahead. "I'm bound to say that I think they should join the Post Carbon Institute," he says. "We're trying to coor-dinate the activities of as many people as we can, as quickly as we can. But it's a tricky proposition. People get very sensitive about this, and well they might. We know that this is not going to be a smooth ride, we definitely know that. But whether they join with us or not, they should educate themselves as much as possible about energy issues so that they are not lured by false hopes. That's one of the most important things, because oth-erwise they will be looking for panaceas and magic-bullet solutions, and I think this will be a grave error."

Another important consideration is that the size and scale of the prob-lems posed by peak oil are simply too great for isolated, individual responses, according to Darley. "Merely changing to compact fluorescent light bulbs is not going to do it," he says. "Of course it's a good idea to use less electricity, and there are many ways of doing it, but only a community response is going to have an effect in a meaningful way." Darley suggests taking the idea of reducing your own energy demand seriously and begin-ning to work with others to do the same on a larger scale. And if you don't know much about energy, he recommends that you consider supporting those who do. "For instance, if you are an electricity user and somebody comes to you with a cooperative wind project proposal, then I suggest you try to look on it favorably," he says. "That's because if you don't like having a wind turbine in your backyard, in my opinion, you've only got one other option—stop using electricity. You should be trying to get your energy, your food, and other essentials as locally as possible, and that means you may have to accept some things that you don't like. Or you can do without."[11]

In February 2006, the Post Carbon Institute launched a new renewable energy initiative, the Local Energy Farm Demonstration Project, in collaboration with the Center for Sustainable Food Systems at the University of British Columbia. This project, part of PCI's relocalization strategy, aims to reduce dramatically and eventually eliminate community dependence on fossil fuels for energy, which would ensure that, after the oil (and gas) peak, public systems delivering electricity, water, sanitation, transportation, food, and many other essential services will continue. Local renewable energy systems would also dramatically reduce CO_2 emissions in communities, helping them to meet their Kyoto commitments and to improve local environmental conditions.[12]

In addition to various energy-crop experiments, the project will include liquid-biofuels production, wind power, PV solar panels, and a biomass digester for methane production, as well as micro-hydroelectricity. PCI hopes to expand this general concept to a larger, community-supported-energy initiative, similar to community-supported agriculture except that community investors would benefit from energy credits or some form of energy-backed currency year-round rather than just receiving locally grown food in the summer and fall. "You have community-supported agriculture to try to help farmers with their costs during the winter and spring when they are not producing very much," Darley says. "Something similar would happen with an energy farm, especially if it was an extension of a food farm. In the early years you would want to attract people who were passionate about it, not those who were simply looking for a quick financial return. Rather, they would be looking for community engagement and energy security. At the moment, energy security is not regarded very highly, but that situation will change soon enough."[13]

For more information on the Energy Farm, visit www.postcarbon .org/ideas/farm.

Willits, California

Although there are many active and dynamic peak oil groups in the PCI Relocalization Network, probably the most successful can be found in Willits, California. Located about 135 miles north of San Francisco, the small rural community of Willits (population 5,098) is a timber town nestled in northern California's coastal mountain range. Four-wheel-drive pickups and logging trucks roll through the streets of Willits, which also claims to host the "Oldest Continuous Rodeo in the State of California" at their annual Fourth of July event. This might seem to be an unlikely location for an active, community response to peak oil, but that's exactly what has been taking place there since the fall of 2004. Sparked by a number of showings of the film *The End of Suburbia: Oil Depletion and the Collapse of the American Dream*, the community has come together to form a comprehensive, organized response to "reinvent the town," in the words of one participant.[14]

"Thirty people showed up the first time," Dr. Jason Bradford, who hosted the film screening, says. There was quite a lot of animated discussion afterwards, and Bradford decided to show the film again. Sixty people turned up the second time, and ninety came to a third showing. The discussions continued at town meetings, the momentum began to build, and soon a coalition of volunteer community members had formed the Willits Economic Localization (WELL) project. The core members of this group subsequently identified fourteen key areas of interest related to the community's survival in a post-peak-oil environment, and then consolidated them into six working groups: food, energy, shelter, water, health and wellness, and social organization.[15]

The project's ad hoc energy group conducted a comprehensive and time-consuming "energy audit" of the city, which provided them with a reference point for present use and a general framework for planning future action. The audit is now contained in a working document produced by WELL and its energy group in October 2005, titled *Recommendations towards Energy Independence for the City of Willits and Surrounding Community*. Early on, the report sets the tone by stressing that "if we want to be able to develop alternative sources of energy in order to maintain some semblance of our society today, we need to do so now while energy is still cheap and plentiful. We cannot afford to wait until fossil fuels decline to the point of severe economic impact—the changes to ensure our survival need to begin today." The energy group has set itself the extremely ambitious goal of complete energy independence for Willits as soon as possible.

The report highlighted the magnitude of the task: it showed that Willits uses more than 1,000-megawatt hours (MWh) of imported energy every day (all units of energy were converted to MWh to make comparisons easier). Energy sources from outside the Willits area include electricity, propane, firewood, natural gas, and gasoline and diesel fuel used for transportation—with gasoline accounting for nearly 50 percent of total imported energy. Industrial uses and home heating, cooling, and cooking make up most of the other half and are heavily dependent on increasingly expensive natural gas and electricity. Of the $30 million that leaves the area annually to pay for this energy, 56 percent is for transportation according to the study. "In addition to seeing how much was being used, one of the advantages of conducting the energy audit was to show how much money was being spent by everyone within the area on energy," says Brian Corzilius, a member of the ad hoc energy group. "In our case, it came to almost 25 percent of after-tax income. That's a lot of money that could be supporting existing local businesses or creating new ones."[16]

Conservation is at the top of a long "to do" list in the draft energy document. "After looking at all the possible local energy produced from a wide range of sources, we realized that we were going to have to reduce our energy consumption by at least 50 percent in order to allow us to even begin to address the issue," Corzelius says. "And that brought us back to the whole idea of economic localization, with the creation of local jobs and local products to replace those from outside the community."

The document also noted that 70 percent of California's electricity is now generated by natural-gas-fired plants, an unsettling fact that adds a note of urgency to the need for the community's move toward energy independence. In response, another key recommendation in the document is the construction of various forms of distributed electrical-generation facilities and the creation of a community-owned utility company. The generation strategies include a wood-waste-fired plant for electricity and heat, possibly

set up as a public-private joint venture; farm-based liquid-biofuel production for biodiesel; biogas produced from biodigesters located at a landfill and
a sewage treatment plant as well as farms; various types of solar electric (PV)
and solar thermal installations on rooftops and hillsides; geothermal heat
pumps; small-scale hydroelectric; and wind turbines. The idea behind the
community-owned utility is that as distributed generation is developed "we
need a way to ensure the local community ultimately gains the benefit and
that revenues generated are controlled and put to work back in the community," the document says.[17] There is another benefit of the community
owned utility approach, according to Corzilius. "If you develop a power generation facility of any type, that also creates local jobs," he says. "Somebody
has to maintain it and monitor it, so each of these projects keeps money
within the community."[18]

TIME FOR ACTION
Once the energy group had completed the initial research phase, it was
time for actual implementation. The energy audit estimated that there is
enough roof space in the city for the production of 25 MWh per day with
photovoltaic panels, which would help to reduce the amount of imported
electricity. The first two community renewable energy projects that the
group is working on involve the addition of solar PV electric systems to the
city's water treatment plant and also to the sewage treatment plant. In a
unanimous vote in January 2006, the Willits city council decided to move
forward on the development of solar electric installations to power operations at these two sites. City council and staff members praised the ad hoc
WELL energy committee for a year of unpaid professional service that provided the technical and financial information needed to help the city make
the decision. City manager Ross Walker said the committee had brought
the technical information to such a level of refinement that city staff didn't
have to put a lot of time into understanding it.[19] The water treatment plant
project may also include a micro-hydroelectrical-generation component,
while the project at the sewage treatment plant might involve the use of
methane-powered microturbines for additional electrical generation.
Additional projects in other locations should follow. However, one
obstacle to moving forward on the PV proposals has been the limited
supply of PV panels. Because of high demand, delays are expected to

continue until production capacity catches up with demand sometime in the next few years.

Funding for these types of initiatives has also been a major obstacle. "One of our biggest challenges is to sort out how we can raise investment capital for large projects," says Bradford. "One way to look at this is that we are a poor, rural area and that we don't have much money to work with. But if you look at what local people are investing in, there probably is a billion dollars in savings bank accounts and there's probably over $5 billion in equities and bonds. Why are people investing in U.S. Treasuries, savings bonds, and the stock market? It has nothing to do with their local community; in fact, it often undermines their local community. So there are actually billions of dollars of potential investment capital that we might be able to shift if we could just convince people to invest in their local community instead. Just getting a portion of the billions that people have invested elsewhere would go a long way toward meeting our local capital needs."[20]

TIME FOR REFLECTION

In March 2006, WELL project participants took time out to take stock of their accomplishments so far, and to do a bit of soul-searching at a two-day weekend retreat. While some in the group felt that the lack of an overall, coherent plan was a problem, others warned against wasting too much time in analysis when so much still needed to be done. "There are tensions that build up, and once in a while you need to come together and just talk through them," Jason Bradford says. "So we had a facilitated process for that, and this led to the decision for us to become a more formal organization. We tried a year of anarchy, and that only gets you so far." The group decided to set priorities, create a strategic plan, explore different levels of official membership, and a number of other issues. The process was extremely productive and led to many valuable insights.

"One of the most important things that a small group like this can do is to understand its role in the community," Bradford explains. "Without understanding your role, it's hard to figure out what you should be doing. So we looked at our role, and the best way to explain it is through the analogy of a midwife. A midwife helps the process of giving birth to something new, but doesn't own the baby. If you think about social networking theory and how things get done, they generally get done through various affinity groups through the city—business organizations, service clubs, and so on. No new group like ours that starts up in town can expect to be able to deal with a problem like peak oil by itself, so knowing what already exists and who your natural allies are and targeting them first is extremely important. It's through building relationships with existing institutions and engaging them in the process of localizing the economy that things get done. So our role really has been to educate the community on where the threats and opportunities are and to be a watchdog for them, and maybe facilitate some projects that help to demonstrate what needs to be done. If a group starts out knowing its role and articulates it up front, a lot of things follow much more easily."[21]

Another major obstacle for the WELL project has been more fundamental, according to Bradford. "One of the biggest challenges is getting people to believe in themselves and to believe in what they can really accomplish as a community working together," he says. "It's possible for us to be self-sufficient in energy and food, but a lot of people don't believe it will ever happen; they think it's impossible or that we just can't do it. The people who have been attracted to the WELL project are the innovative types who have been thinking about these kinds of issues for a while, and they are excited to try it. But what we have to do is get the rest of the community to join in. You have to get the people who actually run the local government, banks, businesses, and so on, to accept this and begin moving toward it, or it's never going to get beyond the wild-eyed dreamers."[22] Fortunately, local officials and business leaders in Willits are beginning to understand not only the importance of the issues involved, but also that the WELL program offers genuine long-term benefits for the entire community, according to Bradford.

"Overall, I think things are going really well," Bradford says. "We started out with a lot of interest from people who are mostly into new ideas and trying something out. What we noticed since then, however, has been a decline of our numbers from those big early meetings. But what has happened is that more of what we want to do is being taken up by existing institutions and individuals embedded in the rest of the community. So now our actual influence is greater, and what's happening is more impressive, even though we don't have as many people showing up to our events. This is a normal diffusion of ideas through the community or a society, and that's the way it has to happen anyway. Understanding your midwife role and how society is transformed helps you to realize that this is okay, and that we are actually gaining ground."[23] For more information on the WELL program, visit www.willitseconomiclocalization.org.

Vermont Peak Oil Network

Although it has not attracted much publicity so far, there has been a dedicated group of people in my home state of Vermont who have been actively forming carbon reduction, peak oil, or relocalization groups in the past few years in a number of cities and towns. But until early in 2006, most of this activity was not well coordinated, and many groups were unaware of one another's existence. That changed dramatically with the creation of the Vermont Peak Oil Network in February 2006, when Annie Dunn Watson from Essex, Vermont, and a number of other peak oil activists met to form the network. Dunn Watson had just returned from a Community Solutions–sponsored training session held in Illinois for community activists. While she was there, she realized that there was already a lot of grassroots, community-based activity going on back in Vermont. "I thought it would really help if we knew about one another's efforts," she says.[24] The seeds for VPON were planted.

VPON now has a comprehensive Web site (www.vtpeakoil.net) that serves as a clearinghouse for information and resources promoting thoughtful, community-based responses to the challenges of peak oil. The site also has a listing and contact information for at least eight active local groups. It also contains links to statewide resources on sustainability, food, farming and gardening, energy, local economy, community building, and transportation. In addition, there is contact info for regional peak oil meet-ups and action groups, as well as a calendar of educational and other special events. A few national links are provided, but the focus is primarily on Vermont.

Although VPON has only been in existence for a short time, Dunn Watson thinks it is already performing a number of valuable services. "I think the primary function is to keep us aware of one another's efforts and to support those efforts wherever possible, and to collaborate and learn from one another," she says. But Dunn Watson sees another important function, human contact, in the face of many challenging issues. "People get charged up by that contact," she says. "In a human, grassroots movement like this, we really benefit by staying in touch and by brainstorming. There is a lot of energy in the room when we are together, and that seems to be another valuable function of the network, keeping people energized and encouraged."

VPON and the peak oil response movement it represents has already been fairly effective on two separate but related tracks. The first, and most obvious, is the activity at the local, community level. These activities tend to differ quite a bit from community to community, depending on particular local needs and interests. "Each place has its own local agenda, which is very important since some communities are very small and rural, while others are larger and more urban," Dunn Watson says. "How each of those groups is reacting provides a model, not only for the other groups, but possibly for the nation, which is going to need to think pretty quickly on its feet about how to react to all of this." The second track is focused on the state legislative level. A number of state representatives have attended various local or VPON meetings, and a useful dialog has begun to develop with them. "I think that this conversation could be very significant in the future," says Dunn Watson. "We are hopeful that this will help to encourage the legislators and other state officials to begin to consider the many challenges that peak oil is going to present."

Overall, Dunn Watson is upbeat about the progress that VPON has made so far. "The network came together in a very short time," she says. "The most important thing is that it exists, and can serve a number of important needs at the same time. It continues to evolve, and that's very exciting."[25]

Addison County Relocalization Network

While VPON has facilitated communication within the peak oil community in Vermont, the real action is at the individual town and city level. There are quite a few communities that have begun to organize active peak oil response programs, including Cabot, Tunbridge, East Montpelier, and Brattleboro, as well as towns in the Mad River Valley, Connecticut River Valley, Chittenden County, and elsewhere. But one of the most ambitious initiatives is in Addison County in the west-central part of the state. Nestled between Lake Champlain to the west and the Green Mountains to the east, this beautiful and largely agricultural part of the state is the home of the Addison County Relocalization Network (ACoRN). After returning to Addison County from the Conference on Peak Oil and Community Solutions in September 2005, I discussed the possibility of forming a local peak oil response group with a number of local area residents. We finally decided to hold a small showing of *The End of Suburbia* to see what would happen.

The screening attracted about a dozen people and sparked a lively discussion that demonstrated quite a bit of interest. A larger meeting was subsequently held at the public library in the town of Middlebury (the main or "shire town" of Addison County) on November 1, where we reviewed and discussed the peak oil response initiative in Willits, California. We decided to officially form a group, and quickly adapted a version of the Willits plan as a template for our own activities to avoid wasting a lot of time trying to figure out what to do. But we also decided to take the Willits plan one dramatic step further, and expanded it to encompass the entire county with its twenty-three small towns and cities. With a total county population of only about 37,000, it's obvious that most of these rural communities are fairly small (generally around 1,000 each) and it just seemed to make more sense to combine our efforts. Over the next few months we organized a steering committee; chose a name; drew up a mission statement; scheduled regular, monthly showings of *The End of Suburbia*; and set up an energy committee and a food committee. A media/education/outreach committee was next, and other committees are planned. "We weren't sure that there would be enough sustained interest to call a second meeting, but the timing was right, and a core group of committed volunteers is all it took to build momentum," says Netaka White, one of the cofounders and a member of the ACoRN steering committee. "Now things are really moving forward."[26]

The ACoRN energy committee has embarked on an ambitious series of initiatives. Following the lead of Willits, our first step was to undertake a comprehensive survey of energy use in the county. We were extremely fortunate to attract a Middlebury College senior who was interested in a challenging January-term project. She undoubtedly got more than she bargained for, but she ended up producing a remarkably professional preliminary study that has provided the basis for several follow-up studies by other Middlebury College students since then. The Middlebury students and several college faculty members also helped establish an extremely productive collaboration between the college and ACoRN that has proven to be mutually beneficial.

The energy study has estimated that over $130 million was spent on energy across the county in 2005, very little of which likely benefited the county's (or even the state's) economy. This rather startling finding has been an invaluable aid as we chart the extent and nature of present energy use and try to develop a plan for a shift to greater reliance on local renewables in the future, which will keep at least some of these energy dollars circulating in the local economy. At the moment, most of Addison County's energy comes mainly from sources outside of the county, the state, and even the nation, leaving area towns extremely vulnerable to price increases and disruptions to supply.

In response to this situation, another part of the energy committee's plans includes the formation of a local, renewable energy cooperative, which will act as a facilitator for, and possibly an investor in, a wide range of renewable energy and conservation initiatives throughout the county. The main focus will be the promotion of locally owned, cooperative, community-supported energy projects as well as a possible retail renewable-energy store location. The committee immediately began to assess the possibility of a small-scale hydroelectric facility at the falls on Otter Creek in downtown Middlebury, but soon learned that there was an existing preliminary plan already underway. ACoRN quickly abandoned its own plans and shifted its support to the existing proposal. The energy committee is also exploring a number of other potential community-supported renewable energy initiatives elsewhere in the county.

Although most of these activities are still rather preliminary, the fact that ACoRN has made this much progress in just over nine months is remarkable. This is primarily owing to a dedicated group of community activists throughout the county who are learning how to work together as an effective team for the common good. To date, ACoRN has over one hundred people on its e-mail list and has representatives from most of the twenty-three towns in the county. "Many of us sense a change is coming," White says. "There is an excitement now about working together to meet the needs of the community while we meet the challenges of the coming decades. We're not minimizing those challenges, but there really is a shared sense of responsibility toward the work that must be done."[27] Expect to hear more from ACoRN in the near future. For additional information about ACoRN visit www.acornvt.org.

The Spiritual Community

No discussion of community would be complete without at least a few words about the spiritual community. The events of the next few decades are going to be an enormous test of our faith, regardless of where that faith is grounded. Our religious and spiritual communities can either play a productive or nonproductive role in these events. It depends. Praying at filling stations for lower gasoline prices is not productive. The need for cooperation between rational people of all faiths—and atheists too—could not be more pressing. If we continue focusing on our differences rather than building on the common core principles at the heart of all the world's great religions, our divided religious communities will only make matters worse at the very time when global cooperation is going to be needed the most.

There are glimmers of hope, however. Truly spiritual people will always find common ground with other faiths. More and more devoutly religious people around the world are beginning to realize that caring for our planetary home is not only in our own best interests, it also honors the creation and the creator, regardless of how he, she, or it may be defined. Some of the most organized and effective initiatives for saving the planet (and its diverse inhabitants) are beginning to spring from a wide variety of religious roots and organizations.

In February 2006, eighty-six evangelical Christian leaders decided to back a major initiative to fight global warming, saying "millions of people could die in this century because of climate change, most of them our poorest global neighbors."[28] Although not all evangelical church leaders supported the initiative, it nevertheless represented a major sea change in thinking in the religious community. The Evangelical Environmental Network, the Baptist General Convention of Texas, Mission Society for United Methodists, the National Religious Partnership for the Environment, and Quaker Earthcare Witness are just a few examples of the groups that have been actively working for better stewardship of the earth. Jewish, Roman Catholic, and Eastern Orthodox leaders also have campaigns underway. And many of these groups are beginning to work together for the common cause. Global problems cross national and religious borders, and so do the solutions. For additional information on this movement, I recommend *Serve God, Save the Planet: A Christian Call to Action* by J. Matthew Sleeth (see bibliography).

Conservation

One final tool that all communities have at their disposal is conservation. While conservation is not a renewable energy strategy per se, it nevertheless is a crucial element of the new energy economy we are about to create. Conservation is the least expensive and least harmful strategy we have available. As noted previously, it can be used to reduce our consumption of a wide range of resources, including but not limited to fossil fuels, electricity, and water. Conservation also reduces the need to build costly new electrical-generation plants and other potentially polluting sources of energy. These reductions can be achieved through a combination of initiatives—including the use of energy-efficient vehicles, appliances, building materials, and other technologies—coupled with imaginative thinking about living better with less. Given the amount of waste in our society, especially in North America, living better with less would be an easy strategy for most people to adopt, except for those at the very bottom of the economic spectrum.

"The further down this road I go, the more supportive of conservation I become," says Piedmont Biofuels' cofounder Lyle Estill. "To me, sustainability and conservation are our only hope; we all have to cut our footprint by a serious amount. The tragedy, of course, is how much fat there is to cut in all of our energy uses; that's the problem." The good news is that the huge amount of waste does provide an opportunity for major improvement in a fairly short time. For those who want to take dramatic action, consumption-reduction strategies such as downshifting and voluntary simplicity offer effective and increasingly popular models. Best of all, conservation can be implemented at the national, state, community, or individual level. You don't have to wait for everybody else; you can get started immediately.

BARRIERS TO PROGRESS

As we have seen, most renewable energy technologies are scalable to suit almost any size project, and consequently there are no real technical barriers to adapting them for community-sized projects. There are, however, a number of nontechnical, mostly political and regulatory barriers that stand in the way of greater acceptance of these community-owned renewable energy projects. Wind energy probably provides the best example of the challenges. There are numerous examples of successful community-owned wind farms in Europe. But in virtually every case, the initial moves by the early wind-power cooperatives were met with stiff opposition from the utilities and regulatory agencies. As noted earlier in the book, utilities almost everywhere resist these kinds of initiatives. Virtually everything is stacked against these local, cooperative ventures. Roadblocks are thrown up against making an affordable or convenient connection. The tariffs (rates) are too low or uncertain to allow the project to obtain financing or operate at a profit. Or production caps for renewable electricity generation are too low to allow a group of people to join in a meaningful group net-metered project, and so on. "The big players are trying to stop this," says Minnesota community wind promoter Dan Juhl. "They say that you can't do enough to make a difference. Well, we beg to differ. Community wind combines long-term, cheap and clean energy, jobs, and environmentally sound policy all into one neat package."[29]

And, as we have also seen, the incentives for large-scale renewable energy projects in the United States are primarily designed for large corporate entities that can benefit from incentives like the production tax credit. Small co-ops generally can't benefit, leading to various convoluted ownership strategies to try to make these smaller projects feasible. Most of these problems could be easily resolved if a European-style feed-in tariff similar to the new Ontario standard offer contracts (advanced renewable tariffs) was enacted across the entire United States. I'm not holding my breath, but this single action would probably do more to ensure greater future electrical energy security for this nation than anything else. All it would take is the (considerable) political vision, courage, and leadership in Washington to do it. "Energy is politics, and the more people that get involved at the economic level the greater the political power that can be wielded," Dan Juhl says. "Minnesota is already pretty progressive on renewable energy; we've been at it for a long time. Now that we are getting so many people involved, it's turning into a real political machine that can drive the policy even further."[30] And that political momentum can be even more effective in dealing with the challenges ahead if it is managed in a collaborative, nonpartisan manner, according to former CIA director James Woolsey. "We should not care whether we are Republicans or Democrats. We shouldn't care whether we like George W. Bush. It does not matter," he says.[31] What does matter is getting something accomplished.

But then there is the not-in-my-backyard issue. An advanced renewable tariff, no matter how well crafted, won't help with this NIMBY problem. I've already mentioned the Cape Wind controversy in Nantucket Sound. But even in my home state of Vermont, I'm sorry to say that this is a major recurring issue. Year after year, a wide variety of public opinion polls consistently show a strong majority of Vermonters support the development of commercial wind farms in the state. A recent poll, conducted in May 2006, revealed that 74 percent of those responding were in favor with only 11 percent opposed (with 15 percent unsure).[32] Yet time after time, when specific wind projects are proposed, a small but vocal group of well-organized opponents often stops the project in its tracks. This anti-wind activity often occurs with the strong support of *The Burlington Free Press*, one of the state's largest newspapers, which has taken a consistently anti-wind editorial stance, giving the impression that most Vermonters oppose wind. But the polls clearly demonstrate otherwise.

Although there are a number of issues, the main arguments voiced by wind opponents center on visual, essentially aesthetic, concerns. I am as concerned about retaining Vermont's attractive rural landscape as anyone, but the time has come for the state's residents to decide whether they want to continue to have electricity or not. If they do, then it's time to begin to take greater responsibility for its production. It's also time to measure these aesthetic concerns against the far more pressing problems of peak oil and global warming, which are a threat to our very survival. Faced with a choice like that, I opt for the wind farms. I think most other rational people would too. The resources to deal with these threats are right here, in our backyards, so to speak, and all we need to do is harness them. No oil wells, no massive oil spills, no environmentally destructive coal mines, no nuclear-disasters-waiting-to-happen, or dependency on unstable or undemocratic foreign regimes are required. What *is* required, however, is a clear understanding of the danger we face, and then the vision, courage, and leadership to respond in a thoughtful and appropriate way. This is going to require some tradeoffs.

"There are a lot of Americans who seem to consider electricity as a God-given right," says Canyon Hydro president Dan New. "On the other hand, there are a lot of people who don't want us to generate electricity with oil, or nuclear, or hydro, or wind. That's the problem. We all need to sit down and agree that we *do* want to see the lights stay on. So that raises the question, how are we going to do that?"[33] Good question.

"Energy isn't easy, and people have to work together to make it happen," says Winooski One's John Warshaw. "Every type of energy has its benefits and its drawbacks. You just have to accept that and decide whether a proposed project, in total, is going to make things better or worse. Wind turbines have their issues, but there's no greenhouse emissions, no coal mining, no radioactive waste, no dams blocking fish travel. So, everything has its plusses and minuses, and I just think we have to move to a place where we have renewable, sustainable, and environmentally acceptable sources of power. We simply have to move away from oil, coal, and nuclear power for many different reasons. People have also got to work together to get past their immediate self-interest in order to get to that point, or we are going to have even more serious problems than we already have."[34]

If the lights (and refrigerators and freezers and furnaces) begin to flicker and go out because we were not able to make these commonsense decisions when we still had the time and ability to do it, I believe most of these wind opponents will wish, in retrospect, that they had a wind turbine in their own backyard. I also suspect that the vast majority of opponents have never actually seen a real wind farm and are essentially frightened by the unknown, and a lot of misinformation. I have seen numerous wind farms, at a distance and up close. Over the years, I've seen hundreds in Europe, Canada, and various states, where folks understand the benefits and are willing to make the modest compromises involved. At first sight, the turbines are admittedly a bit startling. But soon they begin to fade into the background and become part of the scenery, quietly generating local, renewable energy as their blades spin gracefully in the wind. In the nineteenth century, Vermont had hundreds of windmills pumping water for farms all across the state. They were simply part of the working landscape, performing a useful function, and no one seemed to mind. Wind turbines and other renewable energy projects can become a part of the working landscape in Vermont again—and the rest of the nation as well. We're going to need them badly in the very near future, and their presence will be a reassuring and welcome sight in an otherwise uncertain environment.

CHANGING THE MODEL, CHANGING THE DEBATE

Because there has been so much vocal opposition in some locations, it seems to me that this might be a good time for wind supporters, in particular, to step back and think about changing the whole debate by changing the model. This can easily be accomplished by shifting the focus to smaller, carefully sited community-owned projects, especially in areas where opposition to the larger, commercial-scale proposals are running into so much trouble. This same approach would also work for biodiesel or ethanol facilities, and many of the other large-scale renewable energy strategies we've looked at. I am convinced that this simple shift in strategy would defuse much of the opposition, and might convert many NIMBY's into enthusiastic supporters and investors. Obviously, you'll never satisfy everyone, but I think this is an idea whose time has definitely come.

It's mainly a question of addressing people's legitimate concerns and then demonstrating that there is a third—and better—way between giant commercial projects that overwhelm the landscape and the small, backyard wind turbine. This community-based strategy places the turbine (or other renewable energy facility) as close as possible to where it is needed. This reduces or eliminates the need for additional, ugly high-tension power lines, while simultaneously improving the stability of the electricity network. One good-sized wind turbine, for example, could provide all the power needed for a school, business, or manufacturing facility. A small cluster of medium-sized turbines could power a whole community. Add a significant number of rooftop solar panels, small-scale hydroelectric or geoelectric plants, ground-source heat pumps, and a local cooperative biofuels facility or two, and you begin to assemble a picture of greater energy security that provides for a significant proportion of your community's energy needs while generating income, all from local resources. The people employed to operate and maintain these facilities keeps them working (and spending) in their local communities, and eliminates the need for them to commute somewhere else to get to their jobs. The result is energy creation and conservation at the same time. And if the renewable energy facilities power other job-creating activities, such as local manufacture of essential products, you end up boosting the local economy while creating even more jobs. It's a win-win-win proposition.

Finally, there is no question that when pioneering local energy projects like Hull Wind I in Massachusetts or WindShare in Toronto are first proposed, there are a lot of initial concerns and skepticism in the surrounding community. But if the projects are well planned, carefully explained, and properly implemented (as they were in Hull and Toronto), most members of the community eventually come to view the project as a real asset. Then, when an addition to the project is proposed at a later date, the community often approves it much more readily because by that time they understand its many advantages and minimal impacts. This pattern has been repeated over and over again in communities across Europe and increasingly in North America, and clearly demonstrates that most of the initial fears raised prove, in the end, to be groundless.

THE WRONG SOLUTIONS

The growing public recognition and acceptance of the twin dangers of peak oil and global warming in the past year or so has had both a positive and negative impact. On the positive side, the fact that these issues are *finally* being accepted as true and worthy of serious discussion in the United States (virtually the last bastion of denial on the planet) offers hope that something may actually be done to address them. The danger, however, is the potential for pursuing the wrong responses. "My fear is that the whole peak oil concept is getting co-opted by governments and corporations and that they will say, 'Yes, peak oil is here, and *this* is what we need to do about it,'" says The Community Solution's Megan Quinn. "Here on our side of the peak oil movement, we think that could be very dangerous if the 'solutions' involve things like nuclear or coal-fired power plants. It could be disastrous if peak oil is suddenly used to justify all the wrong solutions."[35]

The Post Carbon Institute's Julian Darley agrees. "We don't need to make any predictions about this issue," he says. "Just look at all the hype about the hydrogen economy and the new promotion of the nuclear industry, which are two of the most glaring examples. A lot of people are being sucked in by the nuclear approach, including our political leaders, because it looks like an easy solution to the electricity problem. It's quite clear to me, however, that going down the nuclear road would be catastrophic. It has many dangers, and doesn't do anything at all to solve the liquid-fuel problem unless you turn to the hydrogen economy, which I think is a gruesome trick that has been foisted on people. It's simply daft. It doesn't mean that you can't make hydrogen for a car from solar panels, because you can. But it's one thing to have a few cars and a few hundred panels powering it, and quite another to have a whole national or global system like that. You can make almost anything work in the laboratory, but when you come to do it on a large scale, then the engineering issues are just insurmountable." The other problem with the new hydrogen economy, if you use the California model presently being promoted, is that it uses dwindling supplies of natural gas as the hydrogen source. Wrong solution.

Unlike the new renewable energy industry, which tends to be underfunded and somewhat disorganized, the old energy industry is powerful, deeply entrenched, well financed, and wields enormous political influence. It's galling, indeed, that many of the industries that have strongly resisted or blocked any shift to renewables for decades are presently falling all over themselves to claim that they are suddenly the solution to the very problem they helped to create. Now they are planning on cashing in. This is no coincidence. These industries should *not* be rewarded for their complicity in the crisis we now find ourselves in. It's time for a new model; the old one is broken. It's time we developed our own solutions, even if it means that we have to develop a new economic, political, social, and spiritual model to match. In fact, that's precisely what is needed if we are going to survive the disaster that the old system has created for us. And this new model needs to be bold, creative, dynamic, equitable, and above all else, truly sustainable.

I'm not saying or implying that all corporations are predatory. They're not. There are a growing number of socially responsible companies around the world that are trying to do the right thing, and they have invested heavily in developing strategies that demonstrate their respect for ethical values, people, communities, and the environment. These companies are part of the solution, not part of the problem, and they deserve our support, and our business. Just don't be fooled by all the "greenwash" that is being splashed around by the others who are belatedly trying to make themselves look good.

SECURITY AND OPPORTUNITY

I understand that community-supported energy is a new concept for most Americans, but the potential benefits are enormous. I am convinced that if this idea catches on (and, as we've seen, it's beginning to do so in a number of locations) the entire debate about wind power and other renewables will shift from "not in my backyard" to "*please* in my backyard." Appropriately scaled renewable energy projects integrated thoughtfully into the infrastructure of every community, coupled with conservation, will go a long way toward addressing the energy crisis we currently face, while reversing some of the worst effects of globalization on local economies. The main challenge is to explain the many advantages of this strategy, and then encourage communities to plan, design, and carefully site these projects, and to support them with as much local investment as possible as quickly as possible.

I don't claim to know exactly how all of this is going to play out in the years ahead. There are just too many variables. I *am* certain, however, that we are facing some real challenges to our economy, our political institutions, our social structures, and our security. There's a good deal of rhetoric these days about "homeland security," much of it just a lot of fearmongering on the part of self-serving politicians trying to divide us and make us feel helpless and dependent on their protection. Don't let them fool you. We are *our own* best security. Especially if we are united as a community in our determination to face the future—whatever it may bring—with courage, determination, and compassion. Community-supported energy represents a key element of real *hometown* security and real independence. I firmly believe that a community's commitment to its own future energy security will ultimately prove to be one of the best long-term investments it can possibly make. I hope your community is among the growing number of cities and towns around the world to follow this sensible strategy.

If the global economy collapses (and the national electrical grid along with it) at some point in the future, having a viable source of local energy to power local production of the basic necessities of life will be a community asset of enormous value. But if the economy somehow survives peak oil, global warming, and ecological devastation, these local energy strategies will still make sense. We're going to have to make the transition to renewables sooner or later anyway, so we might as well do it now while we can still afford it. Enhancing the national electrical grid with distributed-generating capacity and strengthening the local businesses in your community will be a bonus, and a gift to the next generation. The opportunities represented by renewables are enormous, and I hear opportunity knocking. Anyone care to open the door?

A lot has happened since I finished the manuscript for this book in mid 2006. Unquestionably, one of the most significant events was the mid-term election in November. As a result of the election, control of both the U.S. House of Representatives and the U.S. Senate shifted to the Democratic Party. This could have profound implications regarding federal energy and climate-change policies. While some pessimists insist that it doesn't matter which party controls Congress, I think the chances for a more focused national renewable-energy effort have been enhanced, and this should also bode well for local renewable-energy initiatives.

Another noteworthy event was the decline in oil prices just prior to the mid-term election. Many voters immediately assumed this decline was politically motivated. While they were probably partly correct, there were a number of other factors at work. As I mentioned in the introduction, oil inventories were at record levels in the U.S. in the summer and fall of 2006, yet prices remained high. This unusual situation was partly due to the prediction that there would be another active hurricane season that might damage the oil and gas infrastructure in and near the Gulf of Mexico. And there was ongoing concern about possible disruptions to oil supplies in other parts of the world. For the most part, these problems did not materialize, but that didn't prevent a good deal of speculation on the international oil markets, which helped to keep oil prices high.

The end of the summer driving season and continuing high oil inventories finally triggered a decline in the price of oil from well over $70 per barrel to under $57. But in the weeks that followed, as the temporary over-supply of oil was wrung out of the market, oil and gas prices began to stabilize, and by early December the price for oil had recovered to over $62. While the price for oil will undoubtedly rise and fall erratically in the coming months and years, it is anticipated that the general trend will be up, especially as we approach the "bumpy plateau" at the top of Hubbert's Peak.

A number of other interesting developments have been reported in recent months. The initial algae-to-biofuel tests by GreenFuel Technologies of Cambridge, Massachusetts, that I mentioned in chapter 6 (Liquid Biofuels) have been successfully completed at the 1,040-megawatt Arizona Public Service Company's gas-fired power plant in Arlington, Arizona. The bioreactors at the site successfully produced feedstock for both biodiesel and ethanol. The next step is to determine whether the process is commercially viable.

On the cellulosic ethanol front, the Broin Companies of Sioux Falls, South Dakota, have announced their intention to build the first cellulosic ethanol biorefinery in the United States at Voyager Ethanol in Emmetsburg, Iowa. Broin plans to convert Voyager from a 50-million-gallon-a-year conventional corn dry-mill facility into a 125-million-gallon-a-year commercial-scale ethanol biorefinery designed to use advanced conversion technologies to produce ethanol from corn fiber and corn stover. The international cellulosic ethanol race is on.

Proposed tidal energy projects in the United States have increased substantially, according to the Federal Energy Regulatory Commission, which lists 38 pending applications for projects in Alaska, Maine, New Hampshire, New York, Oregon, and Washington. In addition to the project in New York's East River mentioned in chapter 4 (Water Power), FERC has issued preliminary permits for tidal projects in the Tacoma Narrows of Washington's Puget Sound, in San Francisco Bay, and eight proposed projects off the coast of Florida. Wave-energy projects are proposed for Narragansett Bay in Rhode Island and Makah Bay in Washington State. Ocean energy seems poised to emerge from the shadows.

Meanwhile, recent British and American studies show that the rate of ice melt in Arctic and Antarctic regions is accelerating well beyond most earlier estimates, prompting concerns about a greater rise in sea levels by the end of the century than previously predicted. This rise could present the threat of substantial flooding, storm surge, and even complete submersion of many of the world's densely populated low-lying areas such as the Nile Delta, Bangladesh, London, and most of Manhattan. NASA scientist James Hansen, one of the most respected American climate researchers, recently warned that the international community has a very narrow window of opportunity—ten years at best—to deal with climate change before global warming moves beyond a critical tipping point.*

If there is any chance of winning this race against time, we must begin taking serious action immediately at the local, state, and national levels to wean ourselves from fossil fuels. So much to do. So little time. Let's get to work now.

December 2006

---

* Speeches, articles, and slide presentations by Hansen are available at www.columbia.edu/~jeh1/.

ORGANIZATIONS AND ONLINE RESOURCES

GENERAL

American Public Power Association
2301 M Street, NW
Washington, DC 20037
Phone: 202-467-2900
E-mail: mrufe@appanet.org
www.appanet.org
APPA is a nonprofit service organization for the nation's more than two thousand community-owned electric utilities that serve more than 43 million Americans.

Earth Policy Institute
1350 Connecticut Avenue, NW
Washington, DC 20036
Phone: 202-496-9290
E-mail: epi@earth-policy.org
www.earth-policy.org
The institute, headed by Lester Brown, is dedicated to providing a vision of an environmentally sustainable economy—an eco-economy—as well as a roadmap of how to get from here to there.

Home Power magazine
P.O. Box 520
Ashland, OR 97520
Phone: 541-512-0201
E-mail: hp@homepower.com
www.homepower.com
This informative magazine is a wonderful resource for small-scale renewable energy strategies in general.

Institute for Local Self-Reliance
927 15th Street, NW, 4th Floor
Washington, DC 20005
Phone: 202-898-1610
E-mail: info@ilsr.org
www.ilsr.org
A nonprofit organization that promotes sustainable, self-reliant communities.

Midwest Renewable Energy Association
7558 Deer Road
Custer, WI 54423
Phone: 715-592-6595
E-mail: info@the-mrea.org
www.the-mrea.org
This nonprofit promotes renewable energy, energy efficiency, and sustainable living through education and demonstration. The site offers information

about events, education and training, discussion forums, and much more.

Minnesotans for Sustainability
www.mnforsustain.org
Although its name includes "Minnesota," this is a huge site, and one of the best and most comprehensive sustainability Web sites anywhere.

Mother Earth News magazine
Ogden Publications, Inc.
1503 SW 42nd Street
Topeka, KS 66609
Phone: 800-234-3368
www.motherearthnews.com
The original guide to living wisely, with articles on green building, alternative energy, organic gardening, natural health, homesteading, whole foods, transportation, and more.

National Religious Partnership for the Environment
49 South Pleasant Street, Suite 301
Amherst, MA 01002
Phone: 413-253-1515
E-mail: nrpe@nrpe.org
www.nrpe.org
An association of independent faith groups across a broad spectrum working for environmental sustainability and justice.

National Renewable Energy Laboratory (NREL)
1617 Cole Boulevard
Golden, CO 80401
Phone: 303-275-3000
E-mail: webmaster@nrel.gov
www.nrel.gov
Comprehensive government site with information on solar, wind, biomass, geothermal, and much more.

Northeast Organic Farming Association (NOFA)
Box 135
Stevenson, CT 06491
Phone: 203-888-5146
E-mail: bduesing@cs.com
www.nofa.org
A nonprofit organization of nearly four thousand farmers, gardeners, and consumers working to promote healthy food, organic farming practices, and a cleaner environment.

Office of Energy Efficiency and
  Renewable Energy
U.S. Department of Energy
www.eere.energy.gov
An excellent government source for renewable energy information.

Rocky Mountain Institute
1739 Snowmass Creek Road
Snowmass, CO 81654
Phone: 970-927-3851
E-mail: outreach@rmi.org
www.rmi.org
A nonprofit organization that fosters the efficient and restorative use of resources to create a more sustainable world.

COMMUNITY

Agraria
www.communitysolution.org/agraria.html
A low-energy-use, small, sustainable community proposed by the Community Solution.

Cohousing Association of the United States
1750 30th Street, #617
Boulder, CO 80301
Phone: 314-754-5828 (voicemail)
E-mail: office@cohousing.org
www.cohousing.org
Building a sustainable society one neighborhood at a time.

Communities Directory
http://directory.ic.org
Online Communities Directory, part of the Intentional Communities Web site, a project of the Fellowship for Intentional Community (FIC). Site contains numerous links as well as a clickable map of communities around the world.

Community Solution, The
P.O. Box 243
Yellow Springs, OH 45387
Phone: 937-767-2161
E-mail: info@communitysolution.org
www.communitysolution.org
A program of Community Service, Inc., dedicated to the development, growth, and enhancement of small local communities. Also the home of the proposed sustainable community of "Agraria."

Earthaven Ecovillage
1025 Camp Elliott Road
Black Mountain, NC 28711
Phone: 828-669-3937
E-mail: info@earthaven.org
www.earthaven.org
An aspiring ecovillage in a mountain forest setting near Asheville, North Carolina, dedicated to caring for people and the earth by learning, living, and demonstrating a holistic, sustainable culture.

Ecocity Builders
P.O. Box 697
Oakland, CA 94604
Phone: 510-444-4508
E-mail: richard@ecocitybuilders.org
www.ecocitybuilders.org
A nonprofit organization dedicated to reshaping cities, towns, and villages for the long-term health of human and natural systems.

Fellowship for Intentional Community
Route 1, Box 155-M
Rutledge, MO 63563
Phone: 660-883-5545
www.ic.org
A nonprofit that provides publications, referrals, support services, and sharing opportunities for a wide range of intentional communities, cohousing groups, ecovillages, community networks, support organizations, and people seeking a home in a community. Publishers of Communities *magazine and the* Journal of Cooperative Living.

PEAK OIL RESPONSE

There are hundreds (if not thousands) of peak oil sites on the Internet. Here are a few of the better ones.

Association for the Study of Peak Oil and Gas
Box 25182
SE-750 25 Uppsala
Sweden
Phone: +46 70 4250604
E-mail: aleklett@tsl.uu.se
www.peakoil.net
ASPO is a network of scientists, affiliated with European institutions and universities, having an interest in determining the date and impact of the peak and decline of the world's production of oil and gas, due to resource constraints.

Energy Bulletin
Peak Oil Primer and links
www.energybulletin.net/primer.php
This is simply one of the best energy sites on the Web, with a very detailed (and nonhysterical) analysis of the present situation and outlook for the future. The home page is updated daily with articles and commentary from around the world on all types of energy. An outstanding resource.

From the Wilderness
655 Washington Street
Ashland, OR 97520
Phone: 541-201-0090
E-mail: service@copvcia.com
www.fromthewilderness.com
This site provides a wealth of information on a wide range of subjects, including but not limited to politics, the economy, regional conflict, and much more.

Life After the Oil Crash
Savinar Publishing
880 Yulupa Avenue, Suite 4
Santa Rosa, CA 95405
E-mail: matt@lifeaftertheoilcrash.net
www.lifeaftertheoilcrash.net
Somewhat opinionated, this site nevertheless contains a lot of useful background information, news and updates, publications, and more on peak oil.

Post Carbon Institute
3683 West Fourth Avenue
Vancouver BC, V6R 1P2
Phone: 604-736-9000
E-mail: info@postcarbon.org
www.postcarbon.org
Excellent site for peak oil response information. Home of the Relocalization Network and the Energy Farm.

The Oil Drum
www.theoildrum.com
An excellent forum for the discussion of peak oil moderated by "a bunch of academics and other folks who think they know what they are talking about. . . . "

Vermont Peak Oil Network
www.vtpeakoil.net
A comprehensive peak oil response site, with a mainly Vermont focus. Click on "Regional Groups" to find a listing for the Addison County Relocalization Network (ACoRN) and other local peak oil response groups around the state.

Willits Economics Localization (WELL)
Willits, CA
www.willitseconomiclocalization.org
One of the oldest and most successful Relocalization Network community peak oil response initiatives in the United States.

BIOMASS

25 x '25
E-mail: info@25x25.org
www.25x25.org
An agriculturally based initiative aimed at producing 25 percent of America's energy from renewable resources by 2025.

Biomass Energy Foundation
E-mail: tombreed@comcast.net
www.woodgas.com
Publishers of a wide range of books on all aspects of biomass energy, but especially on gasification for heat, power, and fuel.

Biomass Energy Resource Center (BERC)
P.O. Box 1611
Montpelier, VT 05601
Phone: 802-223-7770
E-mail: contacts@biomasscenter.org
www.biomasscenter.org
This organization works on projects around the country to install systems that use biomass fuel to produce heat and/or electricity.

Community Power Corporation
8110 Shaffer Parkway, Suite 120
Littleton, CO 80127
Phone: 303-933-3135
E-mail: artsolar@aol.com
www.gocpc.com
A modular biopower system manufacturer.

HearthNet
E-mail: webmaster@hearth.com
www.hearth.com
A comprehensive source of information on wood-, pellet-, coal-, and gas-burning hearth appliances and central heaters with links to manufacturers and local retailers. This site also offers hundreds of articles, thousands of questions and answers, photographs, and much more.

Pellet Fuels Institute
1901 North Moore Street, Suite 600
Arlington, VA 22209
Phone: 703-522-6778
E-mail: pfimail@pelletheat.org
www.pelletheat.org
A trade association that represents the fuel preparation and clean-burning technology of renewable biomass energy resources.

woodheat.org
www.woodheat.org
A noncommercial service in support of responsible home heating with wood. This site offers detailed information on wood-burning technologies, chimneys, firewood, safety, environmental issues, questions and answers, links, and much more.

GEOTHERMAL

Geo-Heat Center
Oregon Institute of Technology
3201 Campus Drive
Klamath Falls, OR 97601
Phone: 541-885-1750
E-mail: geoheat@oit.edu
http://geoheat.oit.edu
Information developed through firsthand experience with hundreds of projects and through extensive research is provided to individuals, organizations, or companies involved in geothermal development.

Geothermal Energy Association
209 Pennsylvania Avenue, SE
Washington, DC 20003
Phone: 202-454-5261
E-mail: daniela@geo-energy.org
www.geoenergy.org
A trade organization that is composed of U.S. companies who support the expanded use of geothermal energy.

Geothermal Heat Pump Consortium
1050 Connecticut Avenue, NW, Suite 1000
Washington, DC 20036
Phone: 202-558-6413
Email: info@ghpc.org
www.geoexchange.org
A nonprofit organization founded in 1994 to promote the growth of energy-efficient, environmentally friendly heating and cooling technology.

Geothermal Research Program
Energy and Geoscience Institute
423 Wakara Way, Suite 300
Salt Lake City, UT 84108
Phone: 801-581-8497
www.egi-geothermal.org
The Geothermal Energy Unit (formerly University of Utah Research Institute) performs basic and applied technical research in geothermal exploration, reservoir delineation, drilling, and production.

Geothermal Resources Council
2001 Second Street, Suite 5
Davis, CA 95616
Phone: 530-758-2360
www.geothermal.org
This nonprofit, formed in 1970, is the primary professional educational association for the international geothermal community.

International Ground Source Heat Pump
Association
374 Cordell South
Oklahoma State University
Stillwater, OK 74078
Phone: 405-744-5175
Email: igshpa@okstate.edu
www.igshpa.okstate.edu
A nonprofit organization that offers a wide range of information on ground-source heat pumps. Their informative Web site also lists certified installers and has numerous links to other sites for additional information.

HYDRO

Canyon Hydro
5500 Blue Heron Lane
Deming, WA 98244
Phone: 360-592-5552
E-mail: info@canyonhydro.com
www.canyonhydro.com
Manufacturer of hydroelectric system components.

Energy Systems and Design
P.O. Box 4557
Sussex NB, E4E 5L7
Canada
Phone: 506-433-3151
E-mail: hydropow@nbnet.nb.ca
www.microhydropower.com
A Canadian manufacturer of micro-hydro systems.

Low Impact Hydropower Institute
34 Providence Street
Portland, ME 04103
Phone: 207-773-8190
E-mail: info@lowimpacthydro.org
www.lowimpacthydro.org
A nonprofit organization dedicated to reducing the impacts of hydropower generation through the certification of environmentally responsible, "low impact" hydropower.

Microhydropower.net
www.microhydropower.net
An informative Internet portal on micro-hydropower in nations around the world.

National Hydropower Association
1 Massachusetts Avenue, NW, Suite 850
Washington, DC 20001
Phone: 202-682-1700
E-mail: help@hydro.org
www.hydro.org
A national trade association dedicated to representing the interests of the hydropower industry.

Ocean Energy Council, Inc.
11985 Southern Boulevard, Suite 155
West Palm Beach, FL 33411
Phone: 561-795-0320
E-mail: oceanenergy@adelphia.net
www.oceanenergycouncil.org
A nonprofit dedicated to ocean energy education and promotion.

Ocean News and Technology magazine
P.O. Box 1096
Palm City, FL 34991
Phone: 772-221-7720
E-mail: techsystems@sprintmail.com
www.ocean-news.com
The only publication of its kind, reporting on the latest ocean industry news, events, and technology developments around the world.

Ocean Power Delivery Limited
104 Commercial Street
Edinburgh EH6 6NF UK
Phone: +44 (0) 131 554 8444
E-mail: enquiries@oceanpd.com
www.oceanpd.com
Scottish manufacturer of a commercial offshore wave energy converter.

LIQUID BIOFUELS

Biodiesel America
www.biodieselamerica.org
A comprehensive site about everything biodiesel, including where to buy it, forums, and much more.

Canadian Renewable Fuels Association
31 Adelaide Street, E
P.O. Box 398
Toronto ON, M5C 2J8
Canada
Phone: 416-304-1324
www.greenfuels.org
A nonprofit that promotes ethanol and biodiesel.

Clean Burn, Inc.
34 Zimmerman Road
Leola, PA 17540
Phone: 717-656-2011
www.cleanburn.com
Manufacturers of industrial waste-oil furnaces and boilers that really work.

Collaborative Biodiesel Tutorial
www.biodieselcommunity.org
This Web site will help you learn how to make your own biodiesel safely, with instructions and advice from experienced producers around the world.

Journey to Forever
www.journeytoforever.org
A highly educational Web site containing a wealth of information about ethanol, biodiesel, and many other sustainable-living subjects, including links to additional sources (in English, Japanese, and Chinese).

Local B 100
www.localb100.com
Home of the Collaborative Biodiesel Tutorial and Biodiesel Homebrew Guide.

National Biodiesel Board
P.O. Box 104888
Jefferson City, MO 65110
Phone: 800-841-5849
www.biodiesel.org
The Web site is an excellent and extensive source of current industry information on biodiesel, news, technical information, links, and much more.

National Ethanol Vehicle Coalition
3216 Emerald Lane, Suite C
Jefferson City, MO 65109
Phone: 573-635-8445
www.e85fuel.com
*Advocates for expanded use of 85 percent ethanol
(E85) motor fuel.*

The Online Distillery Network for Distilleries
  and Fuel Ethanol Plants Worldwide
www.distill.com
*This comprehensive site provides links to the
home pages of distilleries and fuel-ethanol
plants located in various countries around the
world, and to related organizations and other
services.*

Renewable Fuels Association
One Massachusetts Avenue, NW, Suite 820
Washington, DC 20001
Phone: 202-289-3835
E-mail: info@ethanolrfa.org
www.ethanolrfa.org
*The U.S. national trade association for the
ethanol industry, including links, news,
technical reports, and general information.*

Veggie Avenger
www.veggieavenger.com
*A resource for all things biodiesel and straight
vegetable oil.*

Vermont Biofuels Association
www.vermontbiofuels.org
*This nonprofit trade association promotes greater
biofuels production and use in Vermont.*

SOLAR

American Solar Energy Society (ASES)
2400 Central Avenue, Suite G-1
Boulder, CO 80301
Phone: 303-443-3130
E-mail: ases@ases.org
www.ases.org
The ASES publishes Solar Today *magazine,
which often has articles on active solar-heating
systems.*

Cooperative Community Energy (CCEnergy)
534 Fourth Street, Suite C
San Rafael, CA 94901
Phone: 877-228-8700
E-mail: info@ccenergy.com
www.ccenergy.com
*A buyer's cooperative providing quality solar energy
systems to its members at reasonable prices.*

Florida Solar Energy Center (FSEC)
1679 Clearlake Road
Cocoa, FL 32922
Phone: 321-638-1000
E-mail: webmaster@fsec.ucf.edu
www.fsec.ucf.edu
*Provides highly regarded independent third-party
testing and certification of solar hot-water systems
and other solar or energy-efficient products.*

National Aeronautics and Space
  Administration (NASA)
Science Enterprise Program
http://eosweb.larc.nasa.gov/sse/
*Complete solar energy data for the entire planet,
compliments of NASA. Click on "Meteorology
and Solar Energy." After you log in for the first
time, you can gather the information you need
simply by pointing to your location on a world
map. Neat.*

Solar Energy International
P.O. Box 715, 76 S. 2nd Street
Carbondale, CO 81623
Phone: 970-963-8855
E-mail: sei@solarenergy.org
www.solarenergy.org
*This nonprofit educational organization provides
education and technical assistance so that others
will be empowered to use renewable energy
technologies around the world.*

Solar Living Institute
P.O. Box 836
Hopland, CA 95449
Phone: 707-744-2017
Email: isl@rgisl.org
www.solarliving.org
*A nonprofit educational organization that offers
workshops on a wide variety of renewable energy,
energy conservation, and building technologies.*

WIND

Community Wind Collaborative
www.masstech.org/renewableenergy/
Community_Wind/index.htm
*A project of the Massachusetts Technology
Collaborative and the Renewable Energy Trust.*

*Community Wind Financing: A Handbook by the
Environmental Law and Policy Center*
Charles Kubert, et al.
www.elpc.org/documents/WindHandbook
2004.pdf

Danish Wind Industry Association
www.windpower.org/en/core.htm
*Danish wind-turbine manufacturers association site
with latest industry news in Danish and English.*

German Wind Energy Association
www.wind-energie.de/en
*Comprehensive wind-power site in German
and English.*

Hull Wind.org
www.hullwind.org
*Home site for the Hull I and Hull II community-
owned wind turbines in Hull, Massachusetts.*

Minnesotans for an Energy-Efficient Economy
www.me3.org/issues/wind
*An extremely comprehensive site for Minnesota
Wind Energy news and information, with
numerous links for regulations, utilities,
research, and other organizations beyond
Minnesota and around the world. A valuable
research resource.*

Ontario Sustainable Energy Association
401 Richmond Street, W, Suite 401
Toronto ON, M5V 3A8
Canada
Phone: 416-977-4441
E-mail: info@ontario-sea.org
www.ontario-sea.org
*A provincial, nonprofit umbrella organization
formed to implement community sustainable
energy projects across Ontario, Canada. This
site lists workshops, publications, and other
educational materials.*

Wind-Works.org
Paul Gipe
208 South Green Street, #5
Tehachapi, CA 93561
Phone: 661-325-9590
E-mail: pgipe@igc.org
www.wind-works.org
*An excellent online archive of articles and commen-
tary, primarily but not solely, on wind energy.*

WindShare
401 Richmond Street, W, Suite 401
Toronto ON, M5V 3A8
Canada
Phone: 416-977-5093
E-mail: info@windshare.ca
www.windshare.ca
*Home site for the WindShare community-owned
turbine in Toronto, Ontario, Canada.*

Windustry
Great Plains Windustry Project
2105 First Avenue South
Minneapolis, MN 55404
Phone: 612-870-3461
E-mail: info@windustry.com
www.windustry.com
*Simply one of the best and most informative
wind-power sites that promotes wind energy
through outreach, educational materials, and
technical assistance to rural landowners, local
communities and utilities, and state, regional,
and nonprofit collaborations.*

BIODIESEL COOPERATIVES (U.S.)

The Berkeley Biodiesel Collective
Berkeley, California
E-mail: berkeleybiodiesel@yahoo.com
www.berkeleybiodiesel.org

Biofuel Oasis
2465 4th Street at Dwight Way
Berkeley, California
Phone: 510-665-5509
www.biofueloasis.com

The Biofuels Research Cooperative
(Straight Vegetable Oil)
Sebastopol, California
E-mail: veggieoilcoop@yahoo.com
www.vegoilcoop.org

Blue Ridge Biofuels
109 Roberts Street
Asheville, North Carolina
Phone: 828-253-1034
www.blueridgebiofuels.com

Boulder Biodiesel Cooperative
Boulder, Colorado
Phone: 303-449-3277
www.boulderbiodiesel.com

Brevard BioDiesel
Brevard County, Florida
www.brevardbiodiesel.com

Burlington Biodiesel Coop
Burlington, North Carolina
www.burlingtonbiodiesel.org

Corvallis Biodiesel Cooperative
Corvallis, Oregon
www.corvallisbiodiesel.org

Denver Biodiesel
3860 Tennyson Street
Denver, Colorado
Phone: 720-299-1793
www.denverbiodiesel.org

Grease Works! Biodiesel Cooperative
Corvallis, Oregon
E-mail: justin@greaseworks.com
www.greaseworks.org/

Los Angeles Biodiesel Coop
Los Angeles, California
www.biodiesel-coop.org

Olympia Biodiesel
Olympia, Washington
http://home.comcast.net/~olympia_biodiesel

Peninsula Biodiesel Coop
Palo Alto, California
www.peninsulabiodiesel.org

Piedmont Biofuels
P.O. Box 661
Pittsboro, NC 27312
Phone: 919-321-8260
www.biofuels.coop

Portland's Biodiesel Cooperative
Portland, Oregon
E-mail: info@gobiodiesel.com
www.gobiodiesel.com

PrairieFire BioFuels Cooperative
1894 East Washington Avenue
Madison, Wisconsin
Phone: 608-441-5454
www.prairiefirebiofuels.org

Roaring Fork Biodiesel Coop
Roaring Fork, Colorado
www.rfbiodiesel.com

San Francisco Biofuels Cooperative
521 8th Street
San Francisco, California
Phone: 415-267-3998
www.sfbiofuels.org

Tacoma Biodiesel
Tacoma, Washington
www.tacomabiodiesel.org

Utah Biodiesel Cooperative
Park City, Utah
www.utahbiodiesel.org

Yoderville Biodiesel Cooperative
Iowa City, Iowa
www.ybdc.org

ENDNOTES

INTRODUCTION

1 "Algeria: OPEC has reached its production limits," AP, Green Car Congress, March 12, 2005, http://www.greencarcongress.com/2005/03/algeria_opec_ha.html.

2 "OPEC toothless to tame high oil prices: UAE," Reuters, *The Financial Express*, April 7, 2006, http://www.financialexpress.com/fe_full_story.php?content_id=122914.

3 Kevin G. Hall, "Mexican oilfield crucial to U.S. facing decline," MyrtleBeachOnline, March 16, 2006, http://www.myrtlebeachonline.com/mld/myrtlebeachonline/news/nation/14116129.htm (accessed on March 19, 2006).

4 Carl Mortished, "World 'cannot meet oil demand,'" *Times Online*, April 8, 2006, http://business.timesonline.co.uk/article/0,,13130-2124287,00.html.

5 Association for the Study of Peak Oil and Gas, "US DOE has received the warning," February 8, 2005, http://www.peakoil.net/USDOE.html.

6 Julian Darley, *High Noon for Natural Gas: The New Energy Crisis* (White River Junction, VT: Chelsea Green, 2004), 12.

7 "Gazprom owns up to gas shortfall," *Novosti*, Russian News & Information Agency, April 21, 2006, http://en.rian.ru/analysis/20060421/46808586.html.

8 Kelly Levin, Jonathan Pershing, *Climate Science 2005: Major New Discoveries* (World Resources Institute, Washington, DC, 2006), http://climate.wri.org/climate science-pub-4175.html.

9 40MPG.ORG, "Survey: 3 out of 4 Americans fault federal leadership on global warming & alternative energy, back growing state & local efforts," March 15, 2006, http://www.40mpg.org/getinf/031506release.cfm.

10 "D.C. prayer rally to seek lower gas prices," UPI, April 26, 2006, http://www.upi.com/NewsTrack/view.php?StoryID=20060426-114223-5447r.

11 Richard Douthwaite, *Short Circuit: Strengthening Local Economies for Security in an Unstable World* (Dublin, Ireland: Green Books, 1996), 47, 50.

12 Jeffrey J. Brown, "What the mainstream media are not telling you about the run up in oil prices," *Energy Bulletin*, April 20, 2006, http://www.energybulletin.net/15126.html.

CHAPTER 1: OUR ENERGY CHOICES

1 Bloomberg.com, "U.S. economy grows 3.8% for a second straight quarter," Bloomberg, June 29, 2005, http://quote.bloomberg.com/apps/news?pid=10000006&sid=aYexpBFEhf.U&refer=home.

2 U.S. House of Representatives, Committee on Transportation and Infrastructure, Subcommittee on Aviation, "Hearing on commercial jet fuel supply: Impact and cost on the U.S. airline industry," February 15, 2006, http://www.house.gov/transportation/aviation/02-15-06/02-15-06memo.html.

3 James Howard Kunstler, *The Long Emergency: Surviving the End of the Oil Age, Climate Change, and Other Converging Catastrophes of the Twenty-first Century* (New York: Atlantic Monthly Press, 2005), 2–3.

4 Jared Diamond, *Collapse: How Societies Choose to Fail or Succeed* (New York: Viking, 2005), 1.

5   Matthew L. Wald, "Amtrak fires its president in dispute over future," *The New York Times*, November 10, 2005, http://www.commondreams.org/cgi-bin/print.cgi?file=/headlines05/1110-03.htm.

6   *Electric Power Monthly*, January 2004, With Data for October, 2003, Energy Information Administration, Office of Coal, Nuclear, Electric and Alternate Fuels, U.S. Department of Energy, Washington, DC 20585, http://tonto.eia.doe.gov/ftproot/electricity/epm/02260401.pdf.

7   *Natural Gas Annual 2003*, December 2004, Energy Information Administration, Office of Oil and Gas, U.S. Department of Energy, Washington, DC 20585. Table 1, Summary Statistics for Natural Gas in the United States, 1999–2003, http://www.eia.doe.gov/pub/oil_gas/natural_gas/data_publications/natural_gas_annual/historical/2003/pdf/table_001.pdf.

8   *High Noon for Natural Gas*, 20.

9   Ibid, 56–58.

10   Patty Henetz, "A new push for elusive oil shale," *Salt Lake Tribune*, August 9, 2005.

11   *Clean Alternative Fuels, Fischer-Tropsch,* one in a series of fact sheets, March 2002, Environmental Protection Agency, Washington, DC, http://www.epa.gov/otaq/consumer/fuels/altfuels/420f00036.pdf.

12   *High Noon for Natural Gas*, 11.

13   Ibid, 39.

14   Ibid, 59, 60.

15   Ibid, 61.

16   Kari Lydersen, "The false promise of 'clean coal,'" *The New Standard*, March 16, 2006, http://www.alternet.org/envirohealth/33587/.

17   U.S. Department of Energy, "FutureGen—Tomorrow's pollution-free power plant," http://www.fossil.energy.gov/programs/powersystems/futuregen/ (accessed on August 22, 2005).

18   Lyderson, Op. cit.

19   BBC News, "France gets nuclear fusion plant," June 28, 2005, http://news.bbc.co.uk/1/hi/sci/tech/4629239.stm.

20   Bill Sammon, "Massive energy bill signed by the president," *Washington Times*, August 9, 2005, http://washingtontimes.com/national/20050808-093047-7909r.htm.

21   U.S. Department of Energy, "Hydropower resource potential," http://www.eere.energy.gov/windandhydro/hydro_potential.html (accessed on August 23, 2006).

22   In this book "biomass" is used mainly to describe plant matter.

23   U.S. Department of Energy, http://www.eere.energy.gov/RE/bio_fuels.html (accessed on August 26, 2005).

24   Kunstler, op. cit., 115.

25   John G. Edwards, "Conserve energy for peace?" *Las Vegas Review-Journal*, April 12, 2006, http://www.reviewjournal.com/lvrj_home/2006/Apr-12-Wed-2006/business/6820942.html.

26   Kunstler, op. cit., 138–139.

27   Comments on "Changing world technologies" plan to turn garbage into oil, Paul Palmer, Ph.D., chemist, April 9, 2005, http://www.mindfully.org/Technology/2005/Changing-World-Technologies-Palmer9apr05.htm (accessed on August 30, 2005).

28   GreenUniversity.net, "Rocky Mountain Institute: Amory and Hunter Lovins," http://www.greenuniversity.net/Ideas_to_Change_the_World/Lovins.htm (accessed on July 4, 2006).

29 Efficiency Vermont, 2003 Winner, State of Vermont, Government Innovators Network, Harvard University, John F. Kennedy School of Government, http://www.innovations .harvard.edu/awards.html?id=3664.

CHAPTER 2: SOLAR ENERGY

1   A solar tracker is a solar-array mounting rack on a pole that automatically follows the sun during the day to provide the best orientation of the array for maximum output. Track racks can improve electrical output of solar modules by 25 percent or more.

2   The basic solar system design descriptions in this chapter are adapted from *Natural Home Heating: The Complete Guide to Renewable Energy Options* by Greg Pahl (White River Junction, VT: Chelsea Green, 2003).

3   City of Santa Clara Municipal Solar Utility Web site, http://www.ci.santa-clara.ca.us/pub_utility/ws_muni_solar.html (accessed on November 3, 2005).

4   Alan Kutotori, telephone interview by the author, November 14, 2005.

5   Union of Concerned Scientists, "Clean energy, solar water heating," http://www.ucsusa.org/clean_energy/renewable_energy_basics/solar-water-heating.html (accessed on November 13, 2005).

6   Jan TenBruggencate, "Hawai'i residents turning down the juice," *Honolulu Advertiser*, October 15, 2005, http://www.honoluluadvertiser.com/apps/pbcs.dll/article?AID=/ 20051015/NEWS01/510150328/1001/NEWS.

7   Some laboratory prototypes are approaching 40 percent efficiency.

8   Renewable Energy Policy Network for the 21st Century, "Renewable Energy Markets Show Strong Growth, REN21 Releases 'Renewables 2005: Global Status Report,'" http://www.ren21.net/globalstatusreport/g2005.asp.

9   Bill Fleming, telephone interview by the author, October 7, 2005.

10 Ibid.

11 Elena Kann, telephone interview by the author, October 5, 2005.

12 Ibid.

13 CCEnergy, "City of Sebastopol partners with Cooperative Community Energy to launch solar energy program," March 25, 2003, http://www.ccenergy.com/news/press/PR_SolarSebastopol.html.

14 Solar Sebastopol, Photo gallery, http://www.solarsebastopol.com/photos.html (accessed November 13, 2005).

15 Jennifer Bresee and David Room, "Powering down America: Local government's role in the transition to a post-petroleum world," Global Public Media, October 20, 2005, http://www.globalpublicmedia.com/articles/533.

16 *Keeping Current, A Cooperative Community Energy Newsletter*, Summer 2005, "What's happening with Solar Sebastopol?" 4, http://www.ccenergy.com/newsletter/ CCEnergy_Summer2005_Newsletter.pdf.

17 Ibid.

18 Dan Pellegrini, telephone interview by the author, November 21, 2005.

19 This background and description of CCEnergy is adapted, with permission, from the co-op's Web site at http://www.cooperativecommunityenergy.com/faq/index.html (accessed November 20, 2005).

20 Pellegrini, interview.

21 Ibid.

22 Ibid.

23 CCEnergy, "Workshop covered start-to-finish PV system design and installation," June 2002, http://www.ccenergy.com/events/training/MC_article.html.

24 Steve Lyons, telephone interview by the author, November 22, 2005.

25 Johnny Weiss, telephone interview by the author, November 23, 2005.

26 Western Shoshone Defense Project Web site, http://www.wsdp.org/whatsnew.htm#solar0328 (accessed on November 27, 2005).

27 Weiss, interview.

28 Ibid.

29 Soozie Lindbloom, telephone interview by the author, November 23, 2005.

CHAPTER 3: WIND POWER

1 Many people refer to these devices as "windmills," but the correct term is "wind turbine."

2 Patrick Mazza and Eric Heitz, *The New Harvest: Biofuels and Windpower for Rural Revitalization and National Energy Security* (San Francisco: The Energy Foundation, 2005), 36, http://www.ef.org/documents/CompFinalNov18.pdf.

3 Because wind turbines are built (and sold) worldwide, most manufacturers use the metric system to describe the diameter of their turbine's rotors.

4 Most small wind turbines are direct drive and do not contain a gearbox.

5 Paul Gipe, *Wind Power: Renewable Energy for Home, Farm, and Business* (White River Junction, VT: Chelsea Green, 2004), 8, 9.

6 Detailed wind atlases are now available for many areas, making it easier to judge whether sufficient wind can be anticipated in a particular region. Still, nothing beats a year's worth of on-site measurements.

7 Mick Sagrillo, of Sagrillo Power & Light in Forestville, Wisconsin, is a nationally known expert in the wind energy field.

8 Gipe, op. cit., 148.

9 In the U.S., towers are generally sold by the foot and are measured in feet. In most other countries tower heights are measured in meters.

10 Gipe, op. cit., 151–56.

11 Paul Gipe, telephone interview by the author, December 12, 2005.

12 *Public Power: Generating Greener Communities*, Fall 2005, American Public Power Association, 3, http://www.appanet.org/files/PDFs/GenGreenComm.pdf.

13 Paul Gipe, email communication with the author, December 19, 2006.

14 Ibid.

15 Russ Christianson, "Danish wind coops can show us the way," Wind-works.org, http://www.wind-works.org/articles/Russ%20Christianson%20NOW%20Article%201.pdf.

16 Gipe, op. cit.

17 Power-technology.com, http://www.power-technology.com/projects/middelgrunden/ (accessed on December 20, 2005).

18 The Middelgrunden Wind Turbine Co-operative, http://www.middelgrunden.dk/MG_UK/project_info/location.htm (accessed on December 20, 2005).

19 The Middelgrunden Wind Turbine Co-operative, http://www.middelgrunden.dk/MG_UK/wind_cooperative/the_cooperative.htm.

20 "Community wind: The third way."

21 The Ideas Bank, "Samsø to run on 100% renewable energy (project details)," http://ide.idebanken.no/bibliotek_engelsk/ProsjektID.asp?ProsjektID=189.

22 *Wind Power: Renewable Energy for Home, Farm, and Business*, 185.

23 "Danish wind coops can show us the way."

24 Gipe, interview.

25 General Electric Energy, "GE energy supplying wind turbines for new GE project in Germany," September 21, 2005, http://www.gepower.com/about/press/en/2005_press/092105.htm.

26 In the context of electric utility sales, a tariff is a set price, not a tax.

27 Mark Bolinger, "Community wind power ownership schemes in Europe and their relevance to the United States," Lawrence Berkeley National Laboratory, Berkeley, CA, May 2001, 28–30, http://eetd.lbl.gov/ea/EMS/reports/48357.pdf.

28 Ibid, 31.

29 Paul Gipe, "Community-owned wind development in Germany, Denmark, and the Netherlands: Notes on a trip to Northern Europe in 1996," http://www.wind-works.org/articles/Euro96TripReport.html.

30 Bolinger, op. cit., 19.

31 Ibid, 22.

32 Ibid, 25.

33 Baywind Energy Co-operative Limited, "Harlock Hill," http://www.baywind.co.uk/harlockhill.php (accessed on December 23, 2005).

34 Andrew King, "Energy in the UK," *New Sector: Democratic Enterprise and Community Control*, December 2005, 18–19, http://www.newsector.co.uk/PDFs/NS71.pdf.

35 Energy4All, "Westmill wind farm co-op," http://www.energy4all.co.uk/casestudy1.php (accessed on December 24, 2005).

36 David Blittersdorf, telephone interview by the author, December 16, 2005.

37 Database of State Incentives for Renewable Energy (DSIRE), "Vermont Incentives for Renewable Energy," http://www.dsireusa.org/library/includes/incentive2.cfm?Incentive_Code=VT02R&state=VT&CurrentPageID=1 (accessed on December 18, 2005).

38 Blittersdorf, interview.

39 *Renewable Energy; Wind Power's Contribution to Electric Power Generation and Impact on Farms and Rural Communities*, General Accounting Office, September, 2004, 82, 83, www.gao.gov/new.items/d04756.pdf.

40 Blittersdorf, interview.

41 Lisa Daniels, telephone interview by the author, January 3, 2006.

42 One strategy in Minnesota is the so-called "flip" structure under which landowners financially own approximately one percent of their project for the first ten years, the time the PTC is in effect. Investment capital comes from a corporate owner who can use the tax credit. The landowner or co-op earns a small management fee. After ten years, the debts are retired and ownership flips. The landowner or co-op then earns all returns for the remaining life of the turbine, typically about ten years.

43 Dan Juhl, telephone interview by the author, February 16, 2006.

44 "The new harvest," 39, 40.

45 Mark Bolinger, "Community-owned wind power development: The challenge of applying the European model to the United States, and how states are addressing that challenge," presented at Global Windpower 2004, Chicago, Illinois, March 30, 2004, http://repositories.cdlib.org/cgi/viewcontent.cgi?article=2429&context=lbnl.

46 C-BED Community-Based Energy Development, "History of community-based energy development in Minnesota," http://www.c-bed.org/history.html (accessed on February 13, 2006).

47 Juhl, interview.

48 Windustry Fall 2002 Newsletter, "Minwind I & II: Innovative farmer-owned wind projects," http://www.windustry.com/newsletter/2002FallNews.htm.

49 Juhl, interview.

50 Ibid.

51 Renewable Energy Research Laboratory, University of Massachusetts at Amherst, "Wind power on the community scale, community wind case study: Hull," http://www.vma.cape.com/~relweb/Case_Study_Hull_Wind_One.pdf (accessed January 1, 2006).

52 Cape Wind, "Project at a Glance, About the Cape Wind project," http://www.capewind.org/article24.htm (accessed on January 15, 2006).

53 Kevin Dennehy and David Schoetz, "Alliance expands mission," *Cape Cod Times*, August 7, 2005, http://www.capecodonline.com/special/windfarm/allianceexpands7.htm.

54 Amanda Griscom Little, "The wind and the willful," *Grist*, January 12, 2006, http://grist.org/news/muck/2006/01/12/capecod/index.html.

55 Bill McKibben, "No more Mr. Nice Guy," *Grist*, January 12, 2006, http://grist.org/comments/soapbox/2006/01/12/mckibben/index.html.

56 Little, op. cit.

57 Hull Wind, http://www.hullwind.org/ (accessed on January 1, 2006).

58 Hull Wind, "Hull Wind I" (Power Point presentation), May 14, 2005, http://www.hullwind.org/docs/Hull%20Wind%20I.ppt.

59 John Zaremba, "Hulluva windmill: When Hull Wind II starts spinning in March, town will secure its place as an energy visionary," *The Patriot Ledger*, January 14, 2006, http://ledger.southofboston.com/articles/2006/01/14/news/news05.txt.

60 "Community-owned wind power development," 13–15.

61 Kristen Burke, telephone interview by the author, January 5, 2006.

62 Ibid, 12, 13.

63 Electric Power Research Institute, "Giant wind turbines harvest energy for Iowa utilities," June 1999, http://www.epri.com/corporate/discover_epri/news/releases/windfarm.html.

64 Waverly Light and Power, http://wlp.waverlyia.com/search_wind.asp (accessed January 3, 2006).

65 Waverly Light and Power, http://wlp.waverlyia.com/renewable_energy.asp (accessed January 3, 2006).

66 "Project Information," http://www.ci.lamar.co.us/lightpower/info.htm (accessed on January 2, 2006).

67 Traverse City Light & Power, http://www.tclp.org/docs/wind_brochure.pdf (accessed on January 3, 2006).

68 "Wind Turbine Generators," Village of Mackinaw City, http://www.mackinawcity.org/wtg.htm (accessed on January 4, 2006).

69 "Community-owned wind power development," 11.

70 Green Tags, representing the pollution offsets of clean energy generation, are a new way to support renewable facilities that are independently reviewed and endorsed by leading environmental groups.

71 Our Wind Coop, http://www.ourwind.org/windcoop (accessed on January 5, 2006).

72 Jennifer Grove, telephone interview by the author, January 5, 2006.

73 WindShare, "About the Coop," http://www.windshare.ca/about/about_windshare.html (accessed on January 6, 2006).

74 Stewart Russell, telephone interview by the author, January 6, 2006.

75 Ibid.

76 Ontario Sustainable Energy Association, "Ontario takes historic step towards energy future," March 21, 2006, http://www.wind-works.org/FeedLaws/Canada/OSEAHistoricStep.html.

77 Doug Fyfe, telephone interview by the author, January 9, 2006.

78 Ontario Sustainable Energy Association, op. cit.

CHAPTER 4: WATER POWER

1 Mary Bellis, "Lester Allan Pelton, water turbines and the beginnings of hydroelectricity," http://inventors.about.com/library/inventors/bl_lester_pelton.htm (accessed on January 29, 2006).

2 "Winooski One Hydroelectric Project," St. Michael's College, http://personalweb.smcvt.edu/winooskimills/thewinooskiriver/winooskione.html (accessed on January 20, 2006).

3 John Warshow, telephone interview by the author, January 23, 2006.

4 Energy Information Administration, *Electric Power Monthly*, January 2006 Edition, http://www.eia.doe.gov/cneaf/electricity/epm/epm_sum.html.

5 National Hydropower Association, "Water for energy: Country reports: USA," 140, http://www.hydropower.org/downloads/Country%20Reports/USA.pdf.

6 U.S. Department of Energy, "Types of hydropower plants," http://eereweb.ee.doe.gov/windandhydro/hydro_plant_types.html (accessed on January 15, 2006).

7 Canyon Hydro, "Guide to hydropower, part 1: Hydro systems overview," http://www.canyonhydro.com/Resources/Guide/HydroGuide2.htm (accessed on January 16, 2006).

8 Canyon Hydro, "Guide to hydropower, part 1: Hydro systems overview, continued," http://www.canyonhydro.com/Resources/Guide/HydroGuide4.htm (accessed on January 19, 2006).

9 Canyon Hydro, "Guide to hydropower, part 1: Hydro systems overview, continued," http://www.canyonhydro.com/Resources/Guide/HydroGuide5.htm (accessed on January 19, 2006).

10 U.S. Department of Energy, "Hydropower resource potential," http://www.eere.energy.gov/windandhydro/hydro_potential.html (accessed on January 20, 2006).

11 Fred Ayer, telephone interview by the author, January 20, 2006.

12 "Low Impact Hydropower Certification Criteria, Summary of Goals and Standards," http://www.lowimpacthydro.org/documents/criteria_summary.pdf (accessed on February 12, 2006).

13 Ayer, interview.

14 Ibid.

15 Public Utilities Reports, Inc., "The Clark Fork Hydro relicensing: Collaboration or compromise?" June 15, 1999, http://www.pur.com/pubs/3258.cfm.

16 Ayer, interview.

17 Mark Clayton, "A big wave of mini-hydro projects," *Christian Science Monitor*, December 19, 2005, http://www.csmonitor.com/2005/1219/p03s02-sten.html.

18 *Public Power: Generating Greener Communities*, Fall 2005, American Public Power Association, 21, http://www.appanet.org/files/PDFs/GenGreenComm.pdf.

19 Johnny Weiss, telephone interview by the author, January 14, 2006.

20 CADDET, "Small-hydro and municipal water supply," http://www.caddet.org/infostore/display.php?section=1&id=4747 (accessed on January 25, 2006).

21 Carol Ellinghouse, telephone interview by the author, January 27, 2006.

22 Ibid.

23 Will Lindner, "The road less traveled," *Rural Electric Magazine* 60, no. 10 (July 2002), 30–32.

24 Avram Patt, telephone interview by the author, January 23, 2006.

25 U.S. Department of Energy, "How a microhydropower system works," http://www.eere.energy.gov/consumer/your_home/electricity/index.cfm/mytopic=11060 (accessed on January 14, 2006).

26 Paul Cunningham and Barbara Atkinson, "Micro hydro power in the nineties," Energy Systems and Design, http://www.microhydropower.com/staffpubs/staff1.htm.

27 Paul Cunningham, telephone interview by the author, January 31, 2005.

28 Ibid.

29 Ibid.

30 Earthaven Ecovillage, http://www.earthaven.org (accessed on February 2, 2006).

31 Shawn Swartz, telephone interview by the author, February 11, 2006.

32 Ibid.

33 Carolyn Elefant and Sean O'Neill, Ocean Renewable Energy Coalition, "Ocean energy report for 2005," January 9, 2006, http://renewableenergyaccess.com/rea/news/story?id=41396.

34 Dan White, telephone interview by the author, February 8, 2006.

35 Ocean Power Delivery Ltd., "The Pelamis Wave Energy Converter," http://www.oceanpd.com/Pelamis/default.html (accessed on February 9, 2006).

36 Ocean Power Delivery Ltd., "Offshore wave energy," http://www.oceanpd.com/default.html (accessed on February 9, 2006).

37 Wave Dragon, "Technology," http://www.wavedragon.co.uk/technology.html (accessed on March 31, 2006).

38 Nova Scotia Power, "About Us," http://www.nspower.ca/AboutUs/OurBusiness/PowerProduction/HowWeGeneratePower/Hydro.html#ANN (accessed on February 2, 2006).

39 Fujita Research, "Wave and tidal power," July, 2000, http://www.fujitaresearch.com/reports/tidalpower.html.

40 Nova Scotia Power, op. cit.

41 Marine Current Turbines, "World's first offshore tidal current turbine successfully installed," June 16, 2003, http://www.marineturbines.com/mct_text_files/Press%20Release%20MCT%2016%20May%202003%20V2.0.pdf.

42 Marine Current Turbines, "31 May 2005—Seaflow's 2nd birthday," http://www.marineturbines.com/mct_text_files/050531_Seaflow_2nd_birthday.pdf.

43 Marine Current Turbines, "Green light given to next generation tidal energy device," December 19, 2005, http://www.marineturbines.com/mct_text_files/ MCTDTIEHSrelease19Dec05FINAL(2).PDF.

44 Marine Current Turbines, "MCT plans tidal energy farm for North Devon coast," January 27, 2006, http://www.marineturbines.com/mct_text_files/ MCT%20PLANS%20TIDAL%20ENERGY%20FARM%20FOR%20NORTH%20DEVON %20COAST.pdf.

45 The Norwegian company, Hammerfest STRØM AS, claims to be the first to connect a tidal stream turbine to the grid.

46 Verdant Power, "Turbine/Systems Development, East River, New York," http://www.verdantpower.com/initiatives/eastriver.html (accessed on February 5, 2006).

47 National Renewable Energy Laboratory, "What is ocean thermal energy conversion," http://www.nrel.gov/otec/what.html (accessed on February 6, 2006).

48 Dan White, Renewable Energy Access, "Ocean energy: Putting it all in perspective," April 18, 2005, http://www.renewableenergyaccess.com/rea/news/story?id=25108.

49 National Renewable Energy Laboratory, "Ocean Energy Basics," http://www.nrel.gov/learning/re_ocean.html (accessed on February 6, 2006).

50 Renewable Energy Access, "UK organization touts wave and tidal power," January 31, 2006, http://www.renewableenergyaccess.com/rea/news/story?id=42665.

CHAPTER 5: BIOMASS

1 Growing field crops and transport of biomass may generate some CO_2 emissions, depending on the fuel used by the equipment. If biofuels are used, emissions would be minimal.

2 Jeffrey Dukes, "Burning Buried Sunshine: Human Consumption of Ancient Solar Energy," *Climatic Change*, 61 (2003), 31–44, http://globalecology.stanford.edu/DGE/Dukes/ Dukes_ClimChange1.pdf.

3 Andrew Leonard, "The liquid forest," Salon.com, March 17, 2006, http://www.salon.com/ tech/htww/index.html?blog=/tech/htww/2006/03/17/forest_biorefineries/index.html.

4 Robert D. Perlack, Lynn L. Wright, et al., *Biomass as Feedstock for a Bioenergy and Bioproducts Industry: The Technical Feasibility of a Billion-Ton Annual Supply*, U.S. Department of Energy, Washington, DC, 2005, 16, 51, http://www1.eere.energy.gov/ biomass/pdfs/final_billionton_vision_report2.pdf.

5 25x'25 Work Group, "Project Overview," http://www.agenergy.info/index.aspx?mid=28409 (accessed on February 24, 2006).

6 25x'25 Work Group, "25 x '25 Work Group launches three new initiatives," http://www.agenergy.info/index.aspx?ascxid=pagedetail&pid=16487.

7 "Momentum continues to build for 25x'25 campaign," March 23, 2006, http://agenergy.info/docs/QF4967//Newsletter%2003-23-06.pdf.

8 Adapted from *Natural Home Heating: The Complete Guide to Renewable Energy Options* (White River Junction, VT: Chelsea Green, 2003).

9 Ibid.

10 There have been some recent experiments using field crops such as alfalfa as feedstocks for pellets, and this strategy has some interesting potential.

11 *Natural Home Heating.*

12 INFORSE-Europe, "Sustainable energy study tours in Denmark," 2, http://www.inforse.dk/europe/word_docs/study_tours_DK06.doc (accessed on February 20, 2006).

13 My visit to NRG Systems reinforced my decision to install a Tarm boiler in my own home a few months later.

14 David Blittersdorf, interview by the author, May 27, 2005.

15 LEED, Leadership in Energy and Environmental Design, is the nationally accepted standard and rating system for high performance, sustainable (green) buildings developed by the U.S. Green Building Council.

16 Vermont Department of Public Service, Energy Efficiency Division, "The DPS and biomass development," http://publicservice.vermont.gov/energy-efficiency/ee_files/biomass/ee18a.htm (accessed on October 20, 2005).

17 "Schools lock in lower fuel prices," AP, Boston.com, October 16, 2005, http://www.boston.com/news/local/vermont/articles/2005/10/16/schools_lock_in_lower_fuel_prices/.

18 While writing this book I was called for jury duty at the courthouse in the winter heating season of 2005–2006 and found the building to be quite comfortable.

19 David Brynn, telephone interview by the author, March 9, 2006.

20 Only about 20,000 square feet of the community center are occupied and actually heated.

21 Nederland Community Biofuels Project: Technical Description, Operational Results, and Cost Benefits Analysis, Final report evaluating the viability of wood waste as a renewable resource for generating heat and electricity in municipal applications, 2005, iii, http://www.state.co.us/oemc/programs/waste/forest_thinnings/nederland-final_report.pdf (accessed on March 2, 2006).

22 Robert Rizzo, telephone interview by the author, March 3, 2006.

23 Mae-Wan Ho, Peter Bunyard, Seter Saunders, et al., Institute of Science in Society, "Which energy?," 11, 12, http://www.i-sis.org.uk/ISIS_energy_review_exec_sum.pdf (accessed on March 10, 2006).

24 U.S. Department of Energy, "ABCs of biopower, biomass power overview," http://www1.eere.energy.gov/biomass/abcs_biopower.html (accessed on February 19, 2006).

25 *Biomass as Feedstock for a Bioenergy and Bioproducts Industry,* 16.

26 John Irving, telephone interview by the author, February 27, 2006.

27 McNeil Generating Station, http://www.westbioenergy.org/lessons/les04.htm (accessed on February 25, 2006).

28 The permitting for the McNeil Station stipulated that 75 percent of the chips be delivered by railcar in an effort to reduce the potential impact of truck traffic in the Burlington metropolitan area caused by chip hauling tractor-trailers. The rail delivery strategy is more expensive, however.

29 Burlington Electric Department, "Joseph C. McNeil Generating Station," http://www.burlingtonelectric.com/SpecialTopics/Mcneil.htm (accessed on February 25, 2006).

30 Ibid.

31 The New England Power Pool (NEPOOL) is an alliance of approximately 100 utility companies who manage and direct all major energy production and transmission in the New England states.

32 Irving, interview.

33 Michael Valenti, "Preaching to the converted," *Mechanical Engineering* (December 2001), http://www.memagazine.org/backissues/dec01/features/preaching/preaching.html.

34 Ibid.

35 Irving, interview.

36 Ibid.

37 The "Salix Consortium" was formed in New York State (by four principle partners with the cooperation of about 30 private, government, and research institutions) and supported by the U.S. Departments of Energy and Agriculture.

38 Burlington Electric Department, "BED's willow crop experiment," http://www.burlington electric.com/SpecialTopics/willowtree.htm (accessed on February 25, 2006).

39 Noel Vietmeyer, ed., *Producer Gas: Another Fuel for Motor Transport* (Washington, DC: National Academy Press, 1983), 79.

40 Thomas Reed, telephone interview by the author, April 21, 2005.

41 CPC Corporation, "Our Products," http://www.gocpc.com/ (accessed on March 6, 2006).

42 Hydrogen Science Foundation, http://www.hydrogenscience.org/ (accessed March 6, 2006).

43 *Producer Gas: Another Fuel for Motor Transport*, 1–3.

44 Ibid, 51–57.

45 Ibid, 8.

46 Ibid, 84.

47 Reed, interview.

48 Biomass Energy Foundation, "History of Woodgas," http://www.woodgas.com/history.htm (accessed on March 6, 2006).

49 Will Lindner, "A hedge against a wobbly market," *Rural Electric Magazine* 64, no. 3 (December 2005), 28, 29.

50 "Co-op fires up the generators!" *Co-op Currents* 66, no. 5 (August/September 2005), 1, 8, http://www.washingtonco-op.com/news/Aug2005.pdf.

51 Robert Foster, telephone interview by the author, March 1, 2006.

52 Ibid.

53 Central Vermont Public Service, "Cow Power," http://www.cvps.com/cowpower/index.shtml (accessed on February 31, 2006).

54 Central Vermont Public Service, "The ultimate in recycling: Four more farms plan CVPS Cow Power generators, receive grants," April 3, 2006, http://www.cvps.com/cowpower/documents/FourfarmstoproduceCVPSCowPower.pdf.

55 Central Vermont Public Service, "CVPS aims high with cow power," May 3, 2005, http://www.cvps.com/documents/annualmeetingrelease_000.pdf.

56 R. Neal Elliott and Mark Spurr, American Council for an Energy Efficient Economy, "Combined heat and power: Capturing wasted energy," May 1999, http://www.aceee.org/pubs/ie983.htm.

57 Danish Energy Agency, Ministry of Environment and Energy, "Danish Follow-up Programme for Small-Scale Solid Biomass CHP Plants," http://www.dk-teknik.com/services/Air/Combustion/Images/Pdf/Sevilla/Follow%20up%20Programme.pdf.

58 Mikka Kirjavainen et al, "Small-scale biomass CHP technologies, Situation in Finland, Denmark and Sweden," OPET Report 12, VTT Processes and Finnish District Heating Association, 2004, http://www.opet-chp.net/download/wp2/small_scale_biomass_chp_technologies.pdf.

59 "Combined heat and power."

60 EERE Network News, "St. Paul Cogeneration Starts Up a 25-Megawatt Wood-Fired CHP Plant," May 28, 2003, http://www.eere.energy.gov/news/archive.cfm/pubDate=%7Bd%20'2003-05-28'%7D#6359 (accessed on February 28, 2006).

61 Mike Burns, telephone interview by the author, March 3, 2006.

62 Ibid.

63 Tim Franks, "Cows make fuel for biogas train," BBC Newsnight, October 24, 2005, http://news.bbc.co.uk/2/hi/science/nature/4373440.stm (February 19, 2006).

64 Svensk Biogas AB, "Biogas Train, Environmentally Friendly and Cost Effective," http://www.svenskbiogas.se/pdf/en/Biogastrain_produktblad_2005.pdf (accessed on February 20, 2006) (page now discontinued).

65 Franks, op. cit.

66 "Which energy?", 11.

CHAPTER 6: LIQUID BIOFUELS

1 Bill Kovarik, "Henry Ford, Charles Kettering, and the 'Fuel of the future,'" Automotive History Review, no. 32 (Spring 1998), 7–27, reproduced on the Web at http://www.radford.edu/~wkovarik/papers/fuel.html (accessed on March 2, 2006).

2 Greg Pahl, Biodiesel: Growing a New Energy Economy (White River Junction, VT: Chelsea Green, 2005), 23.

3 European Biofuels Group, "A history of biodiesel/biofuels," http://www.eurobg.com/biodiesel_history.html (site now discontinued).

4 Earth Policy Institute, "World ethanol production, 2004," http://www.earth-policy.org/Updates/2005/Update49_data.htm (accessed on March 12, 2006).

5 Earth Policy Institute, "Ethanol's potential: Looking beyond corn," June 29, 2005, http://www.earth-policy.org/Updates/2005/Update49.htm (accessed on March 15, 2006).

6 David Luhnow and Geraldo Samor, "Brazil fills up on ethanol, weans off energy imports," The Wall Street Journal, January 12, 2006, http://www.post-gazette.com/pg/06012/637006.stm.

7 EERE Network News, August 30, 2006, http://www.eere.energy.gov/news/enn.cfm#id_10228 (accessed on August 30, 2006).

8 Mark Clayton, "Carbon cloud over a green fuel," Christian Science Monitor, March 23, 2006, http://www.csmonitor.com/2006/0323/p01s01-sten.html.

9 Philip Brasher, "Ethanol ownership shifting from farmers," DesMoinesRegister.com, March 1, 2006, http://desmoinesregister.com/apps/pbcs.dll/article?AID=/20060301/BUSINESS01/603010350/1029/BUSINESS (accessed on March 15, 2006).

10 David Morris, Institute For Local Self-Reliance, "Ownership matters: Three steps to ensure a biofuels industry that truly benefits rural America," February 2006, http://www.newrules.org/agri/ownershipbiofuels.pdf (accessed on March 15, 2006).

11 National Corn Growers Association, "Ethanol & Coproducts," June 10, 2005, http://www.ncga.com/ethanol/main/production.htm (accessed on March 15, 2006).

12 Joseph DiPardo, Energy Information Agency, "Outlook for biomass ethanol production and demand," http://www.eia.doe.gov/oiaf/analysispaper/biomass.html (accessed on March 21, 2006).

13 Kris Christen, "How green is ethanol as a renewable fuel?" Environmental Science & Technology, February 8, 2006, http://pubs.acs.org/subscribe/journals/esthag-w/2006/feb/policy/kc_ethanol.html.

14 Chippewa Valley Ethanol Company, LLC, "CVEC History," http://www.cvec.com/ (accessed on March 14, 2006).

15 Bill Lee, telephone interview by the author, March 17, 2006.

16 Ibid.

17 David Twiddy, "Energy secretary: Need ethanol sources other than corn," AP, Bellevillenewsdemocrat.com, March 10, 2006, http://www.belleville.com/mld/belleville/news/state/14069334.htm (accessed on March 12, 2006).

18 George Douglas, telephone interview by the author, March 21, 2006.

19 Hemicellulose consists mainly of sugars and sugar acids and can be found in wood or corn fibers. Lignin is the woody cell walls of plant material which is the cementing material between plant cells.

20 Natural Resources Canada, Technologies & Applications, Making Ethanol Fuel, http://www.canren.gc.ca/tech_appl/index.asp?CaId=2&PgId=116 (accessed on March 14, 2006).

21 Douglas, interview.

22 Stuart F. Brown, "Biorefinery breakthrough," *Fortune* (February 6, 2006), 88, http://money.cnn.com/magazines/fortune/fortune_archive/2006/02/06/8367962/index.htm.

23 Iogen Corporation, "Company Profile," http://www.iogen.ca/news_events/company_profile/index.html (accessed on March 14, 2006).

24 Brown, op. cit.

25 Iogen, "Volkswagen, Shell and Iogen to study feasibility of producing cellulose ethanol in Germany," January 8, 2006, http://www.iogen.ca/news_events/press_releases/VW%20Shell%20Jan%2006.pdf.

26 Christopher J. Chipello, "Iogen's milestone: It's selling ethanol made of farm waste," *Wall Street Journal*, April 21, 2004, http://zfacts.com/p/85.html.

27 Author's estimate extrapolated from most recent published figures.

28 "EU Biodiesel Output up 65 Pct in 2005," EBB, Reuters, April 27, 2006, http://www.planetark.com/dailynewsstory.cfm/newsid/36157/story.htm (accessed on April 28, 2006).

29 National Biodiesel Board, "Imperial Western joins elite list of BQ-9000 accredited biodiesel producers," March 8, 2006, http://www.nbb.org/resources/pressreleases/gen/20060310_nbac_imp_western.pdf.

30 "Vietnam firm to make biofuel from catfish fat," Reuters, July 4, 2006, http://www.planetark.com/dailynewsstory.cfm/newsid/37105/story.htm.

31 This description of biodiesel basics and some of the other biodiesel-related material that follows is adapted from my *Biodiesel: Growing a New Energy Economy*.

32 InnovaTek, "InnovaTek Inc. and Seattle BioFuels, Inc. announce the first successful production of hydrogen from 100% biodiesel in a microchannel steam reformer," March 14, 2006, http://www.tekkie.com/news/press_release_3-14-06.htm.

33 Some recent performance results in school bus fleets have indicated an *increase* in fuel efficiency, so the jury is still out on this performance factor.

34 John Sheehan et al., *Life Cycle Inventory of Biodiesel and Petroleum Diesel for Use in an Urban Bus* (Golden, CO: National Renewable Energy Laboratory, 1998), http://www.nrel.gov/docs/legosti/fy98/24089.pdf.

35 In response to a surplus of grains and other crops, the EU established a set-aside program in 1992 that prohibited farmers from growing food or feed crops on 10 percent of their land, while simultaneously allowing them to grow crops such as rapeseed, sunflowers, or soybeans on the set-aside lands for "industrial purposes."

36 "Shell says biofuels from food crops 'morally inappropriate'," Reuters, July 7, 2006, http://www.planetark.org/dailynewsstory.cfm/newsid/37152/story.htm.

37 Mark Clayton, "Algae: Like a breath mint for smokestacks," *Christian Science Monitor*, January 11, 2006, http://www.csmonitor.com/2006/0111/p01s03-sten.html.

38 Maria Alovert, "The grease trap: Co-ops part 1," May 21, 2003, http://lists .subtend.net/pipermail/pdx-biodiesel/2003-May/000721.html.

39 Piedmont Biofuels Cooperative, "Services," http://www.biofuels.coop/services.shtml (accessed on March 11, 2006).

40 Lyle Estill, telephone interview by the author, March 19, 2006.

41 Ibid.

42 This description is of the proprietary Carbo-V process developed by CHOREN Industries of Freiberg, Germany, but could be used, in general, to describe a similar process used by others. CHOREN Industries, "The Key Elements in the Technology: The Carbo-V Process," http://www.choren.com/en/biomass_to_energy/carbo-v_technology/ (accessed on March 23, 2006).

43 *Clean Alternative Fuels, Fischer-Tropsch,* Environmental Protection Agency, March 2002, http://www.epa.gov/otaq/consumer/fuels/altfuels/420f00036.pdf.

44 "Shanghai company starts DME production as trial for transportation," Green Car Congress, April 20, 2006, http://www.greencarcongress.com/dme/index.html (accessed on April 21, 2006).

45 CHOREN Industries, "Questions and Answers," http://www.choren.com/en/faq/ (accessed on March 23, 2006).

46 CHOREN Industries, "Shell partners with Choren in the world's first commercial sunfuel development," August 17, 2005, http://www.choren.com/en/choren/ information_press/press_releases/?nid=55.

47 Environmental and Energy Study Institute, "New bio-refiner with advanced gasification technology coming to the U.S.," Richard Peterson, BCO Newsletter, March 2006, http://www.eesi.org/publications/Newsletters/BCO/bco_31/bco_31.htm.

48 Adapted from *Biodiesel: Growing a New Energy Economy.*

49 "Grant Project Overview," April 10, 2005, http://www.laughingstockfarm.com/Project%20Overview.htm (accessed on April 13, 2006).

50 Ralph Turner, telephone interview by the author, April 13, 2006.

CHAPTER 7: GEOTHERMAL

1 Geothermal Education Office, "Geothermal Energy Facts, Introductory Level," December 23, 2000, http://geothermal.marin.org/pwrheat.html#Q5 (accessed on April 1, 2006).

2 World Bank Group, "Geothermal Energy: Markets," http://www.worldbank.org/html/fpd/energy/geothermal/markets.htm (Google cached version accessed on April 9, 2006).

3 Alyssa Kagel, Geothermal Energy Association, "Geothermal energy 2005 in review, 2006 outlook," January 5, 2006, http://www.renewableenergyaccess.com/rea/news/story?id=41267.

4 Ben Hirschler, "Hydrogen puts Iceland on road to oil-free future," Reuters, May 31, 2002, http://www.planetark.org/dailynewsstory.cfm/newsid/16207/story.htm.

5 U.S. Department of Energy, "A history of geothermal energy in the United States," http://www1.eere.energy.gov/geothermal/history.html (accessed on April 1, 2006).

6 Kagel, op. cit.

7 "Geothermal Energy Facts, Introductory Level."

8 U.S. Department of Energy, op. cit.

9 National Geothermal Collaborative, "Geothermal Direct Use," http://www.geocollaborative.org/publications/Geothermal_Direct_Use.pdf (accessed on April 2, 2006).

10 Brian Brown, P.E., "Klamath Falls geothermal district heating systems," GHC Bulletin (March 1999), 5, http://geoheat.oit.edu/bulletin/bull20-1/art2.pdf (accessed on April 2, 2006).

11 National Geothermal Collaborative, op. cit.

12 Jeff Ball, telephone interview by the author, April 3, 2006.

13 Ball, interview.

14 Michael Milstein, "Energy from the rocks," The Oregonian, June 13, 2001, http://www.oregonlive.com/environment/oregonian/index.ssf?/environment/ oregonian/sc_41geoth13.frame.

15 Ball, interview.

16 Canyon Bloomers, (Formerly M & L Greenhouses), Hagerman, Idaho, Gene Culver, Geo-Heat Center, GHC Bulletin, March 2004, http://geoheat.oit.edu/bulletin/ bull25-1/art7.pdf (accessed on April 3, 2006).

17 Ibid.

18 Kagel, op. cit.

19 U.S. Department of Energy, op. cit.

20 Geothermal Energy Association, "Update on U.S. Geothermal Power Production and Development," March 14, 2006, http://www.geo-energy.org/publications/reports/ 2006%20Update%20on%20US%20Geothermal%20Power%20Production%20and% 20Developmentx.pdf (accessed on April 2, 2006).

21 U.S. Department of Energy, op. cit.

22 Geothermal Energy Association, "US geothermal power poised to pouble, new survey shows," March 15, 2006, http://geo-energy.org//publications/ pressReleases/2006%20US%20Geothermal%20Update%20Release.pdf.

23 U.S. Department of Energy, op. cit.

24 Charles F. Kutscher, Small-Scale Geothermal Power Plant Field Verification Projects, (Golden, CO: National Renewable Energy Laboratory, 2001), 1, www.nrel.gov/geothermal/pdfs/30275.pdf.

25 Charles F. Kutscher, The Status and Future of Geothermal Electric Power, (Golden, CO: National Renewable Energy Laboratory, 2000), 4, www.nrel.gov/geothermal/pdfs/28204.pdf.

26 Ibid, 2, 3.

27 U.S. Department of Energy, op. cit.

28 The Geysers, "History," http://www.geysers.com/history.htm (accessed on April 13, 2006).

29 Kent Robertson, e-mail interview by the author, April 13, 2006.

30 Gerald Nix, telephone interview by the author, April 10, 2006.

31 "About Chena Hot Springs," http://www.yourownpower.com/ (accessed on April 8, 2006).

32 Ibid.

33 Bernie Karl, telephone interview by the author, April 10, 2006.

34 Geothermal Energy Association, "Alaska, Developing Power Plants, Chena,"
    March 2006, http://www.geo-energy.org/information/developing/Alaska/Alaska.asp
    (accessed on April 8, 2006).

35 "Chena Hot Springs Geothermal Power Plant," http://www.yourownpower.com/Power
    (accessed on April 8, 2006).

36 Karl, interview.

37 Ibid.

38 Daniel N. Schochet, *Case Histories of Small Scale Geothermal Power Plants*, proceedings of the
    World Geothermal Congress, Tohoku, Japan, May 28–June 10, 2000, 2201–4,
    http://www.geothermie.de/egec-geothernet/prof/0545.pdf.

39 Some of the material in this section is adapted from my *Natural Home Heating:
    The Complete Guide to Renewable Energy Options*.

40 Kagel, op. cit.

41 Harold Rist II, telephone interview by the author, March 29, 2006.

42 George Hagerty, telephone interview by the author, April 10, 2006.

43 Ibid.

44 GeoThermal, "Luther College Baker Village," http://www.allian
    tenergygeothermal.com/stellent2/groups/public/documents/pub/geo_act_sch_001350.hcsp
    (accessed on April 6, 2006).

45 GeoThermal, "Luther College Center for the Arts,"
    http://www.alliantenergygeothermal.com/stellent2/groups/public/documents/pub/
    geo_act_sch_001351.hcsp (accessed on April 6, 2006).

46 Jerry Johnson, telephone interview by the author, April 7, 2006.

47 Ibid.

48 Todd Chambers, telephone interview by the author, March 30, 2006.

49 Phil Nichols, telephone interview by the author, March 30, 2006.

50 Enwave Energy Corporation, "Fact Sheet,"
    http://www.enwave.com/enwave/view.asp?/dlwc/fact (accessed on April 9, 2006).

51 Enwave Energy Corporation, "Enwave's energy of the future begins cooling Toronto,"
    August 17, 2004, http://www.enwave.com/enwave/news/?s=dlwc&ReleaseID=64.

CHAPTER 8: THE COMMUNITY SOLUTION

1 Author's notes and The Community Solution, "The Second U.S. Conference on Peak Oil
   and Community Solutions," http://www.communitysolution.org/p2conf1.html (accessed on
   April 15, 2006).

2 Ibid.

3 Megan Quinn, telephone interview by the author, April 16, 2006.

4 Ibid.

5 The Community Solution, "Agraria," http://www.communitysolution.org/agraria#conclu
   (accessed on April 16, 2006).

6 Quinn, interview.

7 Celine Rich, telephone interview by the author, April 13, 2006.

8 Post Carbon Institute, http://www.postcarbon.org (accessed on April 14, 2006).

9 Richard Heinberg, *Muse Letter*, no. 160 (August 2005), http://www.museletter.com/
   archive/160.html. (Note: the Oil Depletion Protocol has elsewhere been published as "The
   Rimini Protocol" and "The Uppsala Protocol." All of these documents are essentially identical.)

10 Julian Darley, telephone interview by the author, April 14, 2006.

11 Ibid.

12 Post Carbon Institute, "Local Energy Farm Demonstration Project,"
   http://www.postcarbon.org/ideas/farm/demonstrationproject (accessed on April 20, 2006).

13 Julian Darley, telephone interview by the author, April 20, 2006.

14 R. V. Scheide, "Past the peak: How the small town of Willits plans to beat the coming
   energy crisis," *North Bay Bohemian*, August 10–16, 2005, http://www.metroactive.com/
   papers/sonoma/08.10.05/willits-0532.html.

15 Jason Bradford, telephone interview by the author, October 25, 2005.

16 Brian Corzilius, telephone interview by the author, October 27, 2005.

17 Willits Economic LocaLization (WELL) and Willits Ad-Hoc Energy Group,
   "Recommendations towards energy independence: for the city of Willits and surrounding
   community," October 5, 2005,
   http://www.willitseconomiclocalization.org/EnergyIndependencePlan.pdf (accessed on
   October 23, 2006).

18 Corzilius, interview.

19 Claudia Reed, "Solar city on the way; Council prepares for installation of panels,"
   The Willits News, January 27, 2006, http://www.willitsnews.com/
   Stories/0,1413,253~26908~3216981,00.html.

20 Bradford, interview.

21 Ibid.

22 Ibid.

23 Jason Bradford, telephone interview by the author, April 15, 2006.

24 Annie Dunn Watson, telephone interview by the author, April 26, 2006.

25 Ibid.

26 Netaka White, e-mail interview by the author, April 28, 2006.

27 Ibid.

28 Laurie Goodstein, "Evangelical leaders join global warming initiative," *The New York Times*,
   February 8, 2006, http://www.nytimes.com/2006/02/08/national/08warm.html.

29 Dan Juhl, telephone interview by the author, February 16, 2006.

30 Ibid.

31 John G. Edwards, "Conserve energy for peace?" *Las Vegas Review-Journal*, April 12, 2006,
   http://www.reviewjournal.com/lvrj_home/2006/Apr-12-Wed-2006/business/6820942.html
   (accessed on April 14, 2006).

32 Channel 3 News Poll Results for Friday, May 12, 2006,
   http://www.wcax.com/Global/story.asp?S=4896358.

33 Dan New, telephone interview by the author, January 16, 2006.

34 John Warshaw, telephone interview by the author, January 23, 2006.

35 Quinn, interview.

GLOSSARY

absorber. In passive-solar design, the generally hard, dark, external surface of the heat storage element.

active solar. Systems or strategies that make use of mechanical or other devices to harvest and use solar energy for the production of heat or electricity.

algae. A simple rootless plant that grows in sunlit waters that can be used as a feedstock for biodiesel.

alkyl ester. A generic term for any alcohol-produced vegetable-oil esters or biodiesel.

anaerobic bacteria. Microorganisms that live and reproduce in an environment that does not contain oxygen.

anemometer. An instrument that measures the speed of the wind.

aperture. In passive solar design, the large glass window area that allows the sun to enter the building.

aromatic. A chemical such as benzene, toluene, or xylene that is normally present in exhaust emissions from diesel engines running on petroleum diesel fuel. Aromatic compounds have strong, characteristic odors.

axial-flow turbine. A hydroelectric turbine design in which the shaft through the center of the turbine runs in the same direction as the water flow, much like a boat propeller.

B100 Community. Backyard enthusiasts or "home brewers" of biodiesel in the United States.

bagasse. A fibrous residue of sugarcane stalks that remains after the juice is removed.

barrage. A large dam constructed across a tidal river or estuary used for the generation of electricity from the flowing water of the tides.

baseload generation. A baseload generating facility provides the basic amount of electricity needed year round.

batch process. A method of making biodiesel that relies on a specific, limited amount of inputs for a single batch.

binary power plant. A geothermal power plant that uses a second (binary) fluid with a lower boiling point than water to generate steam to spin the turbine.

biodiesel. A clean-burning fuel made from natural, renewable sources such as new or used vegetable oil or animal fats.

biofuel. A liquid fuel such as ethanol, methanol, or biodiesel, made from biomass resources.

bioheat. A name sometimes applied to biodiesel when it is used for heating purposes.

biomass. Plant material, including wood, vegetation, grains, or agricultural waste, used as a fuel or energy source.

biomass-to-liquid (BTL). A process for converting biomass feedstocks such as straw, wood chips, and other agricultural waste into liquid biofuels.

biopower. Electricity generated by any number of biomass fuels.

breeder reactor. A nuclear reactor that produces as well as consumes fissionable material, especially one that produces more fissionable material.

Btu. British thermal unit(s), a quantitative measure of heat equivalent to the amount of heat required to raise 1 pound of water by 1° Fahrenheit.

bulb turbine. A reaction hydroelectric turbine named for its bulbous shape that also contains the generator.

canola. The common term, especially in North America, for rapeseed.

carbon dioxide (CO₂). A product of combustion and a so-called greenhouse gas that traps the earth's heat and contributes to global warming.

carbon monoxide (CO). A colorless, odorless, lethal gas that is the product of the incomplete combustion of fuels.

catalyst. A substance that, without itself undergoing any permanent chemical change, facilitates or enables a reaction between other substances.

cellulose. A simple sugar and large component of the biomass of plants.

cellulosic ethanol. A type of ethanol produced from the rigid cell wall material that makes up the majority of plants.

cetane number. A measure of the ignition qualities of diesel fuel.

char. The remains of biomass that has been incompletely combusted, such as charcoal if wood is incompletely burned.

chiller. A type of equipment that produces chilled water to cool air.

circulating fluidized-bed. A circulating fuel-combustor system that produces a medium-Btu gas from biomass such as wood chips.

closed-loop system. A geoexchange system using heat from the ground that is collected by means of a continuous loop of piping buried underground that is filled with antifreeze.

cloud point. The point at which diesel and biodiesel fuels appear cloudy because of the formation of wax crystals due to cold temperatures.

coalbed methane. Methane produced from coalbeds in the same way that natural gas is produced from other geological formations.

combined heat and power (CHP, also known as cogeneration). A facility that generates electricity and thermal energy in a single, integrated system.

compression-ignition engine. An engine in which the fuel is ignited by high temperature caused by extreme pressure in the cylinder, rather than by a spark from a spark plug. Diesel engines are compression-ignition engines.

control. In passive solar design, control of summer overheating is provided by roof and cantilever overhangs, awnings, blinds, or even trellises. Other (more active) controls might include thermostats for fans, vents, and other devices that assist or restrict heat flow.

coppice. A traditional management technique utilizing the regrowth from the cut stumps of certain broad-leaved trees.

cord. A measure of a stack of firewood four feet wide, four feet high, and eight feet long.

cordwood. Another name for firewood.

cross-flow turbine. A type of impulse hydroelectric turbine that is not entirely immersed in water, but is used in low-head, high-flow systems.

desuperheater. A special heat exchanger in a geoexchange system used to heat domestic hot water.

dimethyl ether. A synthetic fuel in the form of a colorless, gaseous ether that can be made from coal, natural gas, or biomass.

direct-expansion loop (DX system). A geoexchange system in which the refrigerant runs directly from the heat pump to the underground piping, eliminating one heat exchanger.

direct solar hot water. A solar heating system that directly heats water for domestic hot water or space-heating purposes.

direct use. The use of geothermal hot water for direct heating of pools, greenhouses, fish farms, and other purposes.

distillers grain. A by-product of the alcohol production process.

distributed generation. An energy system in which smaller amounts of power are generated close to where they are used—typically a community, dwelling, or geographical region.

distribution. In both passive and active solar design, the method used to circulate solar heat from the collection and storage locations to other areas of the house.

district heating. A heating system for a group of homes or community that uses a central source of heat, normally distributed via underground pipes.

diversion (or run-of-the-river). A hydroelectric strategy that channels a portion of the river through a canal or pressurized pipe which then spins a turbine connected to a generator, and eventually flows back into the river.

downhole heat exchanger (DHE). Piping in a specially designed well used to extract heat from a geothermal reservoir.

downshifting. A trend to abandon the corporate fast-track and pursue a simpler, slower-paced lifestyle, though not necessarily involving an apreciable reduction in consumption (*see* **voluntary simplicity**).

dry steam reservoir. A geothermal resource that produces steam but very little hot water.

E10 unleaded. Ordinary unleaded gasoline enhanced with ethanol, which is blended at a rate of 10 percent. E10 unleaded is approved for use by every major automaker in the world.

E85: A blend of 85 percent ethanol and 15 percent ordinary unleaded gasoline. This fuel mixture is used in flexible fuel vehicles.

electricity feed-in law (EFL). A law (or tariff) that requires electric utilities to make a connection with a renewable energy producer and to pay a specific long-term fixed price for the power produced.

electrolysis. Using electricity to split molecules, such as water, into positive and negative ions (hydrogen and oxygen ions).

emissions. All substances discharged into the air during combustion.

energy balance ratio. A numerical figure that represents the energy stored in a fuel compared to the total energy required to produce, manufacture, transport, and distribute it.

energy crops. Crops grown specifically for their energy value.

engine-generator set (genset). An internal combustion engine coupled with an electrical generator used to produce electricity.

ethanol. A colorless, flammable liquid that can be produced chemically from ethylene or biologically from the fermentation of various sugars from carbohydrates found in agricultural crops and residues from crops or wood. Also known as ethyl alcohol, alcohol, or grain spirits.

ethyl ester. Biodiesel that is made with the use of ethanol.

exit brine. Hot geothermal water after it has been used for a variety of purposes, often re-injected into the geothermal resource.

feedstock. Any material converted to another form of fuel or an energy product.

Fischer-Tropsch. A process for making a liquid, diesel-like fuel from fossil fuels or biomass.

flash point. The temperature at which a substance will ignite (for biodiesel, above 260°F or 126°C).

flash power plant. A geothermal power plant that uses superheated hot water that "flashes" into steam when brought to the surface.

flexible fuel vehicle (FFV). A car or truck that can run on any blend of unleaded gasoline with up to 85 percent ethanol (E85). A computer in the fuel system automatically compensates for the varying levels of ethanol in the fuel to assure optimum performance at all times.

flow. The amount or volume of water that is available at a hydroelectric site.

fossil fuel. An organic, energy-rich substance formed from the long-buried remains of prehistoric organic life. These fuels are considered nonrenewable, and their use contributes to air pollution and global warming.

gas scrubber. A system or device used to remove impurities from a gas.

gaseous emissions. Substances discharged into the air during combustion, typically including carbon dioxide, carbon monoxide, water vapor, and hydrocarbons.

gasification. A highly efficient process for converting biomass into gas (syngas, wood gas, etc.) through incomplete combustion in a specially designed firebox.

gasifier. A device designed to produce gas (syngas, wood gas, etc.).

gel point. The point at which a liquid fuel gels (changes to the consistency of petroleum jelly) due to extremely low temperature.

geoexchange. The movement of heat energy to and from the earth to heat and cool an indoor environment.

geothermal. Earth heat or heat from the earth—either high-temperature from deep within the earth, or low-temperature from the top fifteen feet of the earth's surface.

gigawatt hour. One gigawatt hour (GWh) equals one million kilowatt hours.

global warming. An increase in the average temperature of the earth's atmosphere, especially a sustained increase sufficient to cause climate change.

glycerin (or glycerol). A thick, sticky substance that is part of the chemical structure of vegetable oils and is a by-product of the transesterification process for making biodiesel. Refined glycerin is often used in the manufacture of soap and pharmaceuticals.

greenhouse effect. The heating of the atmosphere that results from the absorption of reradiated solar radiation by certain gases, especially carbon dioxide and methane.

grid-intertied. An electrical system that is connected to the national electrical system or grid (*see* **off-grid**).

group net metering. Allows farm-based renewable energy systems to send excess power not immediately needed directly back into the electrical grid while crediting the participating farmers (who may have multiple electric meters) for the excess power. This concept may be expanded to cover nonfarm groups of electric customers (*see* **net metering**).

head. The water pressure created by the vertical distance that the water falls from its intake to the turbine in a hydroelectric installation.

heat exchanger. Any device designed to transfer heat.

hemicellulose. Similar to cellulose but less complex, hemicellulose consists mainly of sugars and sugar acids and can be found in wood or corn fibers.

horizontal-axis ("propeller" style). A wind-turbine design in which the main shaft is oriented and spins horizontally (*see* **vertical-axis**).

hot dry rocks (HDR). A very deep geothermal well strategy (largely experimental) designed to make use of high temperatures in rock formations at extreme depths.

hot-water reservoir. A geothermal resource that contains hot water.

hydroelectric plant. An electricity production facility in which the turbine generators are driven by falling water.

hydrokinetic devices. A group of electrical-generation devices that harness the movement of water in the oceans or rivers that do not require the use of dams.

hydrolysis. The decomposition of organic compounds by interaction with water.

hydropower. The use of moving water to provide mechanical or electrical energy.

impoundment. A hydroelectric strategy using a dam to store river water in a reservoir, which is then allowed to flow via a pipe through a turbine attached to a generator, which produces electricity (*see* **diversion** and **pumped storage**).

impulse turbine. A type of hydroelectric turbine driven by one or more high-velocity jets of water that uses nozzles to produce the jets. Examples include Pelton and Turgo.

indirect-injection engine. A (typically older) diesel engine in which the fuel is injected into a prechamber, where it is partly combusted, before it enters the cylinder.

indirect solar hot water. A solar heating system that indirectly heats water for domestic hot water or space-heating purposes through the use of a heat exchanger and antifreeze (*see* **direct solar hot water**).

inverter. An electrical device that converts direct current (DC) to alternating current (AC).

isopentane. A mixture of hydrocarbons extracted from natural gas, sometimes used as the intermediate fluid in a binary geothermal power plant.

Kaplan turbine. A hydroelectric turbine design developed in 1913 by Viktor Kaplan featuring adjustable turbine blades that look like the propeller of a ship.

kerogen. A fossilized material in shale and other sedimentary rock that yields oil when heated.

kilowatt. One thousand watts; ten 100-watt lightbulbs consume a kilowatt of electricity. One kilowatt equals 3,415 Btu.

kilowatt-hour (kWh). A unit of energy equivalent to one kilowatt (1 kW) of power expended for one hour of time.

landfill gas (LFG). All of the gases generated by the decomposition of organic waste in a landfill in the absence of oxygen.

large hydropower. A hydroelectric facility that generates more than 30 megawatts.

lignin. Woody cell walls of plant material which is the cementing material between plant cells.

liquefied natural gas (LNG). Natural gas that has been condensed into a liquid by cooling to an extremely low temperature ($-260°F$ or $-162°C$).

LNG train. The infrastructure, composed of pipelines, terminals, double-hulled cryogenic tankers, and regasification terminals at the receiving end for liquefied natural gas (LNG).

lye. Sodium hydroxide or caustic soda, the same chemical used to unclog kitchen or bathroom drains, that is used as a catalyst in the biodiesel making process.

megawatt. One megawatt equals one million watts.

methane digester. A device, usually a large covered tank, containing organic matter that produces methane gas in the absence of oxygen (*see* **anaerobic bacteria**).

methane hydrates (methane ice). A form of water ice that contains a large amount of methane.

methanol. A volatile, colorless alcohol, originally derived from wood but also from fossil fuels, that is often used as a racing fuel and as a solvent. Also called methyl alcohol.

methyl ester. Biodiesel that is made with the use of methanol.

micro-hydro. A hydroelectric system that generates less than 100 kilowatts.

middle-distillate fuels. Diesel fuel, heating oil, kerosene, jet fuels, and gas turbine engine fuels.

Minnesota flip. A creative legal strategy in which the vast majority of ownership in a wind farm is initially placed in the hands of a large corporate entity which can benefit from the federal production tax credit during the first ten years. Then, after the credit has been fully utilized, the ownership reverts (or flips) back to the actual owners.

munithermal. A geothermal strategy that uses a municipal water system as its water-to-water heat-exchange medium.

net energy balance. The difference between the energy produced and the energy it takes to produce it.

net metering (or net billing). Allows home-based renewable energy systems to send excess power not immediately needed in the home directly back into the electrical grid while crediting the homeowner for the excess power (*see* **group net metering**).

nitrogen oxides (NOx). A product of combustion and a contributing factor in the formation of smog and ozone.

nuclear fission. The fission of uranium 235, an isotope of uranium, that supplies energy for nuclear reactors and atomic bombs.

nuclear fusion. The combining of two small atomic nuclei to form a larger nucleus, sometimes with the release of energy. The use of fusion as an energy source on Earth is still experimental.

ocean thermal energy. A free and renewable energy source caused by the absorbed heat energy of the sun by the oceans.

ocean thermal energy conversion (OTEC). The process or technologies for producing energy by harnessing the temperature differences (thermal gradients) between warm ocean surface waters and that of cold deep waters. Open cycle, closed cycle, and hybrid cycle are different versions of the same general OTEC process.

octane. A rating scale used to grade gasoline for its antiknock properties.

off-grid. An electrical system that is not connected to the national electrical system or grid (*see* **grid-intertied**).

oil shale. A black or dark brown shale containing hydrocarbons that yield petroleum when heated and processed.

open system. A geoexchange system that uses the latent heat in a body of water, which is usually a well but sometimes a pond or stream, as its heat source.

oxygenate. Substances which, when added to gasoline, increase the amount of oxygen in the gasoline blend.

particulate emissions. Substances discharged into the air during combustion. Typically they are fine particles such as carbonaceous soot and various organic molecules.

passive solar. A broad category of solar techniques and strategies for regulating a building's indoor air and (sometimes) water temperatures.

peak load (or demand). The maximum electric load at a specific time.

peak oil. Concerns the long-term rate of petroleum extraction and depletion, but especially the point at which half of the total recoverable resource has been extracted, and production begins to decline.

pellet stove. A heating stove that is fueled with wood pellets.

penstock. A pipe which connects the water source and the turbine in a hydroelectric installation.

petrodiesel. Petroleum-based diesel fuel, usually referred to simply as diesel.

photosynthesis. A process by which plants and other organisms use light to convert carbon dioxide and water into a simple sugar. Photosynthesis provides the basic energy source for almost all organisms.

photovoltaics. PVs or modules that utilize the photovoltaic effect to generate electricity.

pour point. The temperature below which a fuel will not pour. The pour point for biodiesel is higher than that for petrodiesel.

power-purchase agreement (PPA). A contract that defines the selling prices for power and energy, the amount of power and energy sold, and includes provisions to ensure that performance does not fall below a certain standard.

production tax credit (PTC). Part of the 1992 Energy Policy Act. Originally designed to support electricity generated from wind and certain bioenergy resources, the credit provides a 1.8-cent per kilowatt-hour benefit for the first ten years of a facility's operation.

products of combustion. Substances formed during combustion. The products of complete fuel combustion are carbon dioxide and water. Products of incomplete combustion can include carbon monoxide, hydrocarbons, soot, tars, and other substances.

pumped storage. A hydroelectric strategy in which reversible turbine-generators are used to pump water one way (up) and generate electricity when run in the opposite direction (down).

pyrolysis. The chemical alteration of wood, coal, or other combustible materials as a result of the application of heat.

reaction turbine. A hydroelectric turbine that operates fully immersed in water. Examples include Francis, Propeller, and Kaplan.

relocalization. An increasingly popular grassroots strategy for rebuilding the local economy and infrastructure of communities so they may provide the basic necessities of life in an energy-constrained future.

renewable energy. An energy source that renews itself or is renewable from ongoing natural processes.

second-generation biofuel. Biofuels such as cellulosic ethanol and biomass Fischer-Tropsch diesel.

septage. Sediments, water, grease, and scum pumped from a septic tank.

set-aside land. Agricultural lands in the European Union that have been deliberately taken out of food production.

small hydropower. Hydroelectric installations that generate between 100 kilowatts and 30 megawatts.

solar collector array. A group of solar collectors.

solar domestic hot water (SDHW). A (normally) active water-heating system that uses solar energy as the heat source to produce hot water for domestic purposes.

soydiesel. A term used in the United States for biodiesel made from soybean oil.

spark-ignition engine. An internal combustion engine that uses an electrical spark plug to ignite the fuel.

straflo (straight flow) turbine. A hydroelectric turbine design (based on the Kaplan turbine) in which the electric generators are mounted around the outer rim and only the turbine blades are in the water.

straight-veg (straight vegetable oil). The use of filtered, used vegetable oil as a vehicle or heating fuel.

sustainable. Used to describe material or energy sources that, if carefully managed, will provide at current levels indefinitely.

switchgrass. A summer perennial grass that is native to North America that is resistant to many pests and plant diseases, and capable of producing high yields with very low applications of fertilizer, making it an ideal biomass feedstock.

synthesis gas (also known as syngas, producer gas, biosyngas, gengas, or wood gas). A synthetic gas produced through the gasification of biomass.

synthetic oil. Oil manufactured using the Fischer-Tropsch process which converts carbon dioxide, carbon monoxide, and methane into liquid hydrocarbons of various forms.

tar sands. A mixture of tar (or bitumen), sand, and clay.

tars. The volatile products of incomplete combustion.

thermal depolymerization (TDP). A process for converting biomass and other materials into oil.

thermal mass (or heat sink). The heat storage element in a passive solar design.

tidal energy. Energy from ocean tides caused by the changes in gravitational forces exerted by the moon, and to a lesser extent, the sun.

tidal (or marine) turbine. A turbine design that is similar to wind turbines, except the blades are underwater and spin in response to the tides.

tight sands gas. Conventional natural gas extracted from unconventional reservoirs.

tipping point. In peak oil, the tipping point is reached when global demand for oil finally exceeds the supply, causing exponential increases in the price for oil.

ton. The most common heat-pump sizing measurement; a one-ton heat pump can generate 12,000 Btu of cooling per hour at an outdoor temperature of 95°F, or 12,000 Btu heat output at 47°F.

transesterification. A chemical process that uses an alcohol to react with the triglycerides contained in vegetable oils and animal fats to produce biodiesel and glycerin.

triglycerides. Fats composed of three fatty-acid chains linked to a glycerol molecule.

truss tower. A tower for a wind turbine that normally uses a framework of interconnected steel members in its design.

tubular tower. A tower for a wind turbine made from gently tapered sections of steel tubing.

turbine. A rotary engine, usually containing a series of vanes, blades, or buckets mounted on a central rotating shaft that is actuated by a current or stream of water, steam, or air.

unitherm. A group geoexchange system used to heat and cool a number of buildings from the same source.

vertical-axis ("egg-beater" style). A wind turbine design in which the main shaft is oriented and spins vertically (*see* **horizontal-axis**).

viscosity. The ability of a liquid to flow. A high-viscosity liquid flows slowly, while a low-viscosity liquid flows quickly.

voluntary simplicity. A trend to abandon the corporate fast-track and pursue a simpler, slower-paced lifestyle that involves significant reduction of both income and consumption.

waste vegetable oil (WVO). Used vegetable oil (normally from restaurants) that can be used as a feedstock for biodiesel, or burned directly as fuel in a retrofitted vehicle or heater.

water turbine. A rotary engine containing a series of blades or buckets mounted on a central rotating shaft that is actuated by a current or stream of water.

wave energy. A free and sustainable energy resource created as wind blows over the ocean surface.

wind farm. A group of wind turbines.

wind turbine. A turbine for the generation of electricity that is powered by the wind.

wood alcohol. Produced from the destructive distillation of wood, this highly toxic type of alcohol contains methyl alcohol as its main ingredient.

wood pellet. A fuel for heating normally made from highly compressed sawdust.

yellow grease. A term used in the United States to refer to recycled cooking oils.

BIBLIOGRAPHY

Alovert, Maria. *Biodiesel Homebrew Guide*. 10th ed. San Francisco: n.p., 2005. Available from www.localb100.com/book.html.

Boyle, Godfrey. *Renewable Energy*. 2nd ed. New York: Oxford University Press, 2004.

Brown, Lester R. *Plan B 2.0: Rescuing a Planet under Stress and a Civilization in Trouble*. New York: W. W. Norton & Company, 2006.

Chiras, Dan. *The Solar House: Passive Heating and Cooling*. White River Junction, VT: Chelsea Green, 2002.

Darley, Julian. *High Noon for Natural Gas: The New Energy Crisis*. White River Junction, VT: Chelsea Green, 2004.

Darley, Julian, David Room, and Celine Rich. *Relocalize Now!: Getting Ready for Climate Change and the End of Cheap Oil*. Gabriola Island, BC, Canada: New Society Publishers, 2006.

Davis, Scott. *Microhydro: Clean Power from Water*. Gabriola Island, BC, Canada: New Society Publishers, 2003.

Deffeyes, Kenneth S. *Beyond Oil: The View From Hubbert's Peak*. New York: Hill & Wang, 2005.

———. *Hubbert's Peak: The Impending World Oil Shortage*. Princeton: Princeton University Press, 2001.

Douthwaite, Richard. *Short Circuit: Strengthening Local Economies for Security in an Unstable World*. Dublin, Ireland: Green Books, 1996. (This classic is unfortunately out of print and hard to find; an online version containing the complete original text is available at www.feasta.org/documents/shortcircuit/contents.html).

Doxon, Lynn Ellen. *The Alcohol Fuel Handbook*. Haverford, PA: Infinity, 2001.

The End of Suburbia: Oil Depletion and the Collapse of the American Dream. Gregory Green, 78 min. The Electric Wallpaper Company, 2004. [videocassette or DVD]

Estill, Lyle. *Biodiesel Power: The Passion, the People, and the Politics of the Next Renewable Fuel*. Gabriola Island, BC, Canada: New Society Publishers, 2005.

Gipe, Paul. *Wind Power: Renewable Energy for Home, Farm, and Business*. White River Junction, VT: Chelsea Green, 2004.

Harvey, Adam, Andy Brown, Priyantha Hettiarachi and Allen Inversin. *Micro-Hydro Design Manual: A Guide to Small-Scale Water Power Schemes*. London: ITDG Publishing, 1993.

Heinberg, Richard. *Powerdown: Options and Actions for a Post-Carbon World*. Gabriola Island, BC, Canada: New Society Publishers, 2004.

———. *The Party's Over: Oil, War and the Fate of Industrial Societies*. Gabriola Island, BC, Canada: New Society Publishers, 2003.

Holmgren, David. *Permaculture: Principles and Pathways Beyond Sustainability*. Hepburn, Victoria, Australia: Holmgren Design Services, 2002.

Ikerd, John. *Sustainable Capitalism*. Bloomfield, CT: Kumarian Press, 2005.

Kachadorian, James. *The Passive Solar House*. Rev. ed. White River Junction, VT: Chelsea Green, 2006.

Komp, Richard J. *Practical Photovoltaics: Electricity from Solar Cells*. 3rd ed. Ann Arbor, MI: Aatec Publications, 2001.

Kunstler, James Howard. *The Long Emergency: Surviving the End of the Oil Age, Climate Change, and Other Converging Catastrophes of the Twenty-first Century*. New York: Atlantic Monthly Press, 2005.

Langley, Billy C. *Heat Pump Technology*. Englewood Cliffs, NJ: Prentice Hall, 2001.

Pahl, Greg. *Biodiesel: Growing a New Energy Economy*. White River Junction, VT: Chelsea Green, 2005.

————. *Natural Home Heating: The Complete Guide to Renewable Energy Options*. White River Junction, VT: Chelsea Green, 2003.

————. *The Complete Idiot's Guide to Saving the Environment*. Indianapolis, IN: Alpha Books, 2001.

Potts, Michael. *The New Independent Home: People and Houses that Harvest the Sun, Wind, and Water*. White River Junction, VT: Chelsea Green, 1999.

Schumacher, E. F. *Small is Beautiful: Economics as if People Mattered*. New York: Harper Perennial, 1989.

Simmons, Matthew. *Twilight in the Desert: The Coming Saudi Oil Shock and the World Economy*. Hoboken, NJ: Wiley, 2005.

Sleeth, J. Matthew. *Serve God, Save the Planet*. White River Junction, VT: Chelsea Green, 2006.

Solar Energy International. *Photovoltaics: Design and Installation Manual*. Gabriola Island, BC, Canada: New Society Publishers, 2004.

Thomas, Dirk. *The Woodburner's Companion: Practical Ways of Heating With Wood*. 2nd ed. Chambersburg, PA: Alan C. Hood, 2004.

Tickell, Joshua. *From the Fryer to the Fuel Tank: The Complete Guide to Using Vegetable Oil as an Alternative Fuel*. 3rd ed. New Orleans: Tickell Media Productions, 2003.

INDEX

A Glossary is provided starting on page 327.

25 x '25 Energy Work Group, 147–148

absorbers, 36–37
AC generators, 113
active solar. *See also* photovoltaic (PV)
  systems
    direct liquid systems, 40–41
    indirect liquid systems, 40–41
    installations, 42–43
    phase-change liquids, 41
    solar domestic hot water (SDHW),
      37–41
    space heating, 41–42
active solar energy systems, 20
Addison County Relocalization Network
  (ACoRN), 283–286
Agraria (planned intentional community),
  270, 272
agriculture. *See also* biofuels; biomass;
  greenhouses
    and Community Supported Agriculture
      (CSA), 267
    and cooperatives, 80
    feedstock crops, 190
    Laughing Stock Farm, Maine, 220, 222,
      223
    natural gas effects on, 4, 12
    wind power and, 98
air-source heat pumps, 243
The Alcohol Fuel Handbook (Doxon, L. E.),
  200
algae and biodiesel feedstock, 208
Alovert, Maria "Mark", 209, 211
Alternate Energy Revolving Loan Program
  (AERLP), 95
anaerobic bacteria, 160, 177
anaerobic digesters, 160–161
Andrews, Steve, 260
anemometer, 66
anti-wind activists, 289
aperture, 36–37
Arkansas River Power Authority (ARPA),
  96
arrays, 44

Arsene d'Arsonval, Jacques, 139
ASTM (D-6751), 212
Avista, 116–117
axial-flow turbines, 135
Ayer, Fred, 114, 118

B100 fuel terminal, 212
backup heaters, 39
backyard cooperatives, 210
backyard ethanol, 199–200
bacteria growth, 205
Ball, Jeff, 229–231
barrages, 134–136
barriers to progress, 287–291
Basalt Middle School, 60
batteries, 125
battery backup, 47
Battle Mountain Band of Te-Moak
  Western Shoshone, 59
Bay of Fundy, 135
Baywind Energy Cooperative, 82
Berkeley Biodiesel Collective, 211
Bernstein, Barry, 177
big box stores, 5
binary power plants, 235
biodiesel
    Berkeley Biodiesel Collective, 211
    characteristics of, 203–204
    development and history of, 187
    energy balance of, 207
    feedstock issues, 207–210
    fundamentals of, 201–203
    National Biodiesel Board, 210
    Northeast Biodiesel plant, 212
    operational issues of, 204–205
    production of, 23–24, 200
    statewide biodiesel requirement, 194
    strategy summary, 265
    uses of, 188, 205–206
biodiesel communities, 208–210
biodiesel cooperatives, 210–215
biodiesel feedstock and algae, 208
Biodiesel: Growing a New Energy Economy
  (Pahl, G.), 187

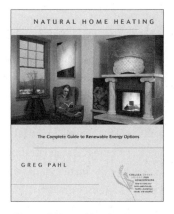
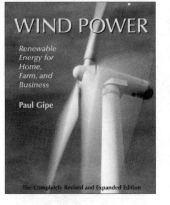
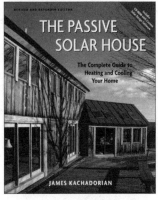

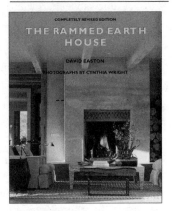
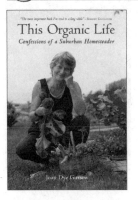